Toilet

NYU SERIES IN SOCIAL AND CULTURAL ANALYSIS
General Editor: Andrew Ross

Nice Work If You Can Get It: Life and Labor in Precarious Times
Andrew Ross

City Folk:
English Country Dance and the Politics of the Folk in Modern America
Daniel J. Walkowitz

Toilet: Public Restrooms and the Politics of Sharing
Edited by Harvey Molotch and Laura Norén

Toilet

Public Restrooms and
the Politics of Sharing

EDITED BY
Harvey Molotch and Laura Norén

NEW YORK UNIVERSITY PRESS
New York and London

IN MEMORIAM

Mary Douglas
&
Lenny Bruce

NEW YORK UNIVERSITY PRESS
New York and London
www.nyupress.org

References to Internet websites (URLs) were accurate at the time of writing.
Neither the author nor New York University Press is responsible for URLs
that may have expired or changed since the manuscript was prepared.

Library of Congress Cataloging-in-Publication Data
Toilet : public restrooms and the politics of sharing /
edited by Harvey Molotch and Laura Norén.
p. cm. — (NYU series in social and cultural analysis)
Includes bibliographical references and index.
ISBN 978–0–8147–9588–0 (cl : alk. paper) — ISBN 978–0–8147–9589–7
(pb :alk. paper) — ISBN 978–0–8147–6120–5 (e-book : alk. paper)
1. Toilets. 2. Toilets—Social aspects.
I. Molotch, Harvey Luskin. II. Norén, Laura.
GT476.T65   2010
392.3'6—dc22   2010018796

New York University Press books are printed on acid-free paper,
and their binding materials are chosen for strength and durability.
We strive to use environmentally responsible suppliers and materials
to the greatest extent possible in publishing our books.

Manufactured in the United States of America

c 10 9 8 7 6 5 4 3 2 1
p 10 9 8 7 6 5 4 3 2 1

Contents

Acknowledgments

FOR FINANCIAL SUPPORT leading to this book, we thank Catharine Stimpson, professor and dean of the Graduate School of Arts and Science, New York University. We also gratefully acknowledge funding and active participation of the New York University Office of Planning and Design and its leader, Lori Pavese Mazor, associate vice president. Further financial assistance came from the Department of Sociology, the Department of Social and Cultural Analysis and its programs in Metropolitan Studies and Gender and Sexuality, and the NYU Center for the Study of Gender and Sexuality. The originating conference, "Outing the Water Closet," held November 2007 at the Center for Architecture, received financial support as well from the AIA (New York Chapter) and its director, Rick Bell. Beth Stryker provided programming support. Harvey Molotch thanks the Russell Sage Foundation, where he was a visiting scholar during the completion of this book.

Both editors offer profuse appreciation to Ilene Kalish, editor extraordinaire. We also thank Simon S. Lawrence for photographic assistance. Judith Stacey, ardent supporter from the outset, read and critiqued the entire manuscript.

Introduction

Learning from the Loo

Harvey Molotch

PUBLIC AND TOILET do not sit well together. The discord goes beyond words. Using the facility—let's call it that for now—involves intensely private acts. Focusing on the public restroom, as the contributors to this book make it their business to do, thus opens a tense domain. But it is a route worth taking, precisely because of the shadow under which it normally falls. By going there, we have the potential to make discoveries with implications for personal hygiene, psychological stress, and social betterment. We can also learn about power and the capacity to shape others' life chances. Hence a group of scholars, drawn from the diverse disciplines of sociology, anthropology, law, architecture, archaeology, history, gender studies, and cultural studies, conjoin to face the facts, unpleasant or otherwise, of the loo.

Even the home bathroom can unleash embarrassment, shame, or criticism when family members detect by sight, sound, or scent what one another are up to. Places such as restaurants or shopping centers introduce anonymity (often welcome) but also concerns about having to share intimate space with people whose intrusions may make us anxious and from whom we want to keep our intimacies separate. The person in the next stall may be the boss or a rival co-worker. The open-to-all facility, as in a public park or train station, invites its own range of anxieties—a person of filth or stranger ready to attack.

So here we have the problem at hand: the toilet involves doing the private in public and under conditions only loosely under the control of the actors involved. By using this tension as springboard, we open up larger issues of what people think they need to protect, how they go about securing that protection, and who succeeds and who does not. We examine the forces that organize such accomplishment and failure—how neighborhoods, cities, cultures, and nations provide for some and not for others. Put bluntly, peeing is political, and so is taking a shit and washing up. We use the word *toilet* inclusively, calling on its French connotation, to cover people's acts of intimate caring to keep themselves decently competent and without bodily offense.

The toilet is a foundational start point where each of us deals directly with our bodies and confronts whatever it provides, often on a schedule not of our own making. The animal in us comes to the fore, and we must accommodate to its tendencies and demands. It is "bare life," as it surfaces in social existence.[1] When we are away from home, we must use some variant of public provision to civilize and prepare for the social world to follow. When on the road, it becomes the ultimate "backstage" of life (in Erving Goffman's famous term)—where we set up our "presentation of self." And when we are readying that performance, it becomes truly important who knows what we are up to and just how they know it. It also matters what precisely we have to work with when we prepare. Without adequacy in these regards, we are almost literally nothing in this world.

There are the material practicalities. How far away is the facility? Is it clean and clean in the sense that matters to me? Do I have access by right? By money? By force? Will there be a proper Western toilet on which I can sit, or will I have to squat? If I am from a squatting part of the world, must I risk physical contact with a public appliance? Toilet paper must be present for an American or European. For an Indian in India, water through a wash pipe that can be directed toward anus or vulva is the utter necessity. Will there be paper covers I can put on the toilet seat, or must I—as women often do in the United States—hover over the seat rather than make the physical contact? And if I lack the muscles to hover, will the waste deposited by prior hoverers with poor aim have been cleaned away by some others (and just whom?) or be a basis for subsequent filth and cringe-worthy horror?

Even within the rich parts of the world, toilet suffering occurs. People in poor neighborhoods have fewer places to go, in part due to lower density of restaurants, bars, and shops, as well as of public restrooms. Besides picking and choosing to whom courtesy will be extended, commercial establishments are not always open, and indeed those in poor neighborhoods commonly have irregular hours.[2] The law in most U.S. states requires that all commercial establishments have toilet facilities and that these facilities must be open to anyone on the premises—whether employee, customer, or even bystander. Most people do not know such laws exist or the procedures for their enforcement. The result is nonenforcement and more problems for people who need a toilet.[3]

Businesses that do provide restrooms, out of respect for either the law or human need, hence end up doing more than their share. In New York City, Starbucks has been called "the city's bathroom." According to one study of Manhattan Starbucks restroom use, the great majority who go into the restroom are not customers; they come and go without buying anything.[4] Company policy gives Starbucks managers discretion over whom they will allow to use the restroom. Sometimes, especially if there is no Starbucks nearby, people must fall back on privilege or special cunning to find an appropriate spot. Those with reasons to fear official authorities have special motive to avoid anything that might be seen as confrontation or stepping out of place. And Starbucks does not, in fact, saturate U.S. cities, being absent from many deprived neighborhoods altogether. Lack of access affects homeless people most directly (and they are not welcome at Starbucks), not only because they routinely confront the humiliation of soiling themselves but also because without a place to wash up, their smell or surface dirt marks them off as offensive.

On a much more massive scale, hundreds of millions of people in poor countries of the world lack even rudimentary toilet access (in either public or private facilities). In India, the proportion of the population that defecates in the open is about half, according to WHO-UNICEF statistics.[5] This produces bacteriological and parasitic infections as human waste mixes into the public environment and in water sources. Where such services are available only in a few places and at certain times of day, people alter their lives in strong ways to deal with the mayhem that going out might otherwise entail. Schoolgirls in India, for example (as well as women workers) say that they go the whole day without "elimination," rather than face the danger, embarrassment, or dirt of the communal facility.[6] They report they "have a system" in which they restrict their intake to affect what they eliminate. The anxieties of basic bodily needs impact the ability to gain an education or livelihood. For many people in the world, when they do find a place to go, it is out in the open or in a ditch that runs with untreated sewage, creating a massive public health problem for those downstream—in some sense, for the majority of people on earth. Speaking of the most crucial of the defilements, Rose George put it deftly when she observed in her trenchant book *The Big Necessity* that "the irony of defecation is that it is a solitary business yet its repercussions are plural and public."[7] A nongovernmental organization called the World Toilet Organization (WTO), based in Singapore, indeed does all it can to foster interest and spread information about the need for and best approaches to dealing with the toilet. Within India, a most remarkable organization, Sulabh International, operates a far-flung system of projects to enhance construction of facilities, to educate, and to press public officials to remediate. Highly aware of the interdependence of modes of excreta with social justice, it has as one of its goals to "remove the practice of untouchability and social discrimination and to restore human rights and dignity of the persons who clean human excreta manually . . . and to bring them in the mainstream of society, so that they could live at par with other castes, which was the dream of Mahatma Gandhi."[8]

Whatever the setting or scale of the problem, we have in the toilet an instrument and institution that both reflects how people and societies operate and also reinforces the existing pattern. Precisely because the toilet operates somewhat in hiding, those who plan, manage, and control its use often act on their own, without a public to which they must provide detailed and explicit accounts of what they are doing. The toilet thus operates irresponsibly. Compared to other artifacts, arrangements, and patterns of usage, it thus resists change—however unjust, damaging, or inefficient things may be. The mechanisms of the flush, the imperatives of access, or the pleasures and punishments of elimination do not arise in annual corporate reports (not even of plumbing companies), much less in a U.S. presidential campaign. The

UN General Assembly did declare the year 2008 the International Year of Sanitation, but the publicity and action such a declaration was meant to raise around the world fell far short of the mark. Major world leaders, before or since, do not meet to solve the problems or make them—life-and-death matters for poor countries—the basis of major speeches. So Americans and people in other rich countries do not have such unpleasantness on their radar. And even in this book, centered as it is on the United States and similar societies, problems of basic sanitation do not much arise. But some of the issues, such as silence and gender, do overlap for both rich and poor places. Perhaps taking them up in a less desperate context—it can be a hope—will facilitate the larger discourse.

Gender Runs through It

In rich places or poor, and more than anywhere else in public life, toilets inscribe and reinforce gender difference. The markings are for "Men" or "Women." There is not just difference but also hierarchy, given that women must wait in their separate lines, whereas men usually do not have to wait at all. This "great binary," much less the inequality with which it is often associated, is neither natural nor inevitable. The contributors to this book raise alternative possibilities, both as cultural reformulations and as architectural alternatives.

The toilet allows us to ask what precisely it might mean to provide equality, an issue that becomes complex if it is granted that groups include individuals who are different in essential ways. Men and women do have different biological characteristics, and those imply different types and degrees of spatial need. Besides different body "plumbing" that affects, in particular, the discharge of urine, women menstruate. This creates more visits, longer stays, and higher stakes for creating and managing a mess. Women need special trash receptacles for their "sanitary napkins" and tampons. And there are cultural differences between men and women, including how they think about the toilet and what they take to be how they should behave in regard to it. There are also different types of clothing and different grooming practices to contend with—lipstick and long hair more commonly for women than for men. Maybe women socialize over the sinks in ways that men do not.

Are these also traits that should be acknowledged and designed into the equality metric by giving women, for example, more space than men? Or should women be encouraged to adhere more closely to men's practices? If we pressure women to change their ways, such as spending less time primping or maybe even peeing standing up, wouldn't that alleviate the problem? Maybe the solution rests in a new appliance, combined with new habits. Various devices exist that a woman places over her vulva to direct the bladder's contents into a container that can then be emptied into a sink, toilet, or urinal.[9] Under names such as She-Pee, they have existed for generations. Another solution is a female urinal that, placed lower on the wall than men's and with the use of the hand to help direct the flow, enables women to urinate in a semistanding position. Finally, women could pee in a crouched position, as is common in Africa, which would probably encourage wearing skirts or sarongs without underpants.

Especially from a male perspective, these may all seem appropriate requests to make of women. But if women are not taken to be the "other," the special case, but instead we start from their perspective, things are seen differently. If we are to respect women as they now exist in the social milieu in which they indeed find themselves, then solutions require increasing provisions for them to use. To provide equality for individual men and individual women, unequal resources must be distributed to the two groups—a proper rationale for affirmative action of any sort.[10] This brings us to a principle more generally relevant for dividing resources so that disadvantaged people share in life opportunities on an equitable basis. In practice, this pans out— in the toilet instance—to at least a two-to-one ratio favoring the size of the women's restroom compared to the men's, sometimes three to one. Women with elaborate hair styles, to take one mundane example, need more counter time than do women with simple cuts. The proper ratio cannot be universal because gendered cultural practices of dress and sociability do differ among subgroups within nations and certainly across the world. We need a calculus based on both habit and respect to come up with the right ratios. So we need to acknowledge not only gender difference but also cultural variations in those gender relations. As a more general matter, how tailor-made should facilities be—sacrificing low-cost efficiencies of standardization for treating each group of people in the manner that most befits their sensibilities and needs?

We could solve the problem, at least of unequal gender access, by ending separation, full stop. Such has been the argument that some gender-rights groups have put forward—including those representing persons whose identities simply do not follow the binary. Where do transgender people go? Or people whose biology renders them intersexed? But what does recognizing their solution then take away from others? What about men who already suffer from paruresis ("pee shy"), which they fear will be exacerbated by the presence of women in the vicinity? There are women who use the women's room as a respite from male supervision, a place where "the girls" can let their hair down and exercise solidarity. Some women report that the ladies' room is where they learned as girls how to do their hair, hold their bodies, use menstrual products, and adjust their clothes—with pals and relatives fussing around them in real time.[11]

Bodies differ not just by gender (and its own variations) but also by age, circumstance, and accidents of fate—imposing various degrees of ability and disability made evident through the public toilet. We were all children once, a life stage of special needs, with respect not only to supervision and care but also to size and proportion—being able to climb onto the toilet seat, hold oneself on without falling in, and reach for the soap, faucet, or paper towels. Elderly people encounter a different battery of challenges, in part from the need to go often but also because frailty weakens the ability to manipulate equipment or improvise around broken or inadequate appliances. And some people just run into trouble along the life-course way, from either a ski accident, a car crash, or a debilitating ailment. Will there be grab bars for hoisting above the seat? For those who are blind, will towels and soap be in the usual spots that can be located precisely because of their standardization? People using wheelchairs need doors light enough to give way without causing the wheelchair to roll backward when the doors are pulled open—not the "high-quality" heavy doors that are found, perhaps ironically, in some deluxe circumstances.[12]

Governance

The personal is political, something that rings especially true in the present case. Much of governance is about sorting, such as where you have to live to vote or how old you must be. Age similarly determines who can drive a car, drink alcohol, or be in porn movies. There is also an array of appropriate locational juxtapositions: special privilege goes to proximity for mothers and babies, such as at border crossings and security gates (more so than for fathers). At the toilet, social pollution can occur with the wrong mixing—boys and girls at a certain age and, depending on cultural moment and context, rich and poor, white and black, business and coach. And the precise details come into play: men and women can be physically near (utterly adjacent, in fact), so long as there is sight separation, for example, by a thin wall between stalls marked for men and for women. Legislating, by physical arrangement or legal authority, arises from this need to enforce the specifics of who will be with whom, where and when.

The restroom thus becomes a tool for figuring out just how a society functions—what it values, how it separates people from one another, and the kinds of trade-offs that come to be made. As is often the case with rules meant for collectivities, some people are left out or feel themselves ignored by the classifications or classified according to criteria that are for them irrelevant or even noxious. Until the 2000 count, the U.S. Census form carried the implicit insistence that people come in races, with each person being one race or another. Mixed-race people were something other. The census now allows people to check off as many races as they wish, but the result still forces choices to be made among categories that some may regard as nonsense. So it is with the gendered restroom, and hence there are protests, sometimes organized, to reorganize the distinctions. Transgender people may protest, as now they sometimes do on university campuses, aiming for the "restroom revolution" to remove the problematic of where they belong.

Even among those who agree that public toilets are important and that there should be more of them, trade-offs must be made. It is easy to see why these allocations and affordances are taken to be every-one's business, even those who almost never use a public restroom. As citizens and taxpayers, what goes on is being done "in our name." However intimate the goings-on, people assume responsibility for others' actions. They feel a need to set up conditions that will do least evil and most good and encourage the right thing—however defined. Citizens become "moral entrepreneurs."[13]

What should be facilitated and what hindered? One important goal of industrial designers, and often of architects as well, is to serve people's actual practices. When conjuring up a new product, design-ers thus study, often in minute detail, exactly how people open a can or exactly what they do with their telephone. They then design to enhance these practices, sometimes in ways not much envisioned in the original product. To pursue this design principle to its logical ex-treme, perhaps bathrooms should accommodate drug users' need for shelf space or provide a spigot tailored to flush out a syringe. Diabet-ics would also gain. Perhaps the stalls would be outfitted with stations for clean needle exchange, as U.S. restrooms now provide for condom acquisition?

Or is it public responsibility to do all possible to inhibit acts that are deemed deleterious or that otherwise interfere with public effi-ciency? Besides drugs, people use toilets to nap, have sex with others or themselves, read, write graffiti, vandalize, chat (in person or on cell phone), groom, smoke, or nip a drink. They hide from teachers and parents. Sociologists go in to write up their notes.[14] Shoplifters use the stalls to change into stolen clothes; others just steal a moment of solace.[15] And at least once in a while, a violent crime gets commit-ted. What types of control—signs of warning, guards at the entrance, video surveillance in the stalls—are appropriate, and how far should they go?

At the extreme, officials often close down the facility altogether rather than put up with the bad acts—thus depriving everyone of access. Those utterly without alternatives then excrete in public spaces—yielding visual and olfactory ugliness, among other con-sequences. Such patterns of prohibition, exclusion, adaptation, and befoulment raise more general issues of how to respond to disliked behaviors. It brings home the problem of social control: What price are we willing to pay to limit activities about which we might disapprove? How much does potential offense of the few color our imagination, politics, and resource expenditure? To what extent does a society design operations and governing procedures out of fear of the miscreants, versus adding satisfaction to collective needs?

The Physical and the Social

Social scientists as well as architects and designers look for ways to understand how the social and the physical relate. A consensus seems to emerge: both "count"—the physical world determines behavior and social patterns, but the social also affects the physical world. In the most radical formulations, the physical and the social are not separate things at all; they operate as a single synthetic force.[16] So how does it happen? Toilets can help clarify.

Galen Cranz points out in her instructive book *The Chair* that because we use chairs (and toilets, of course, also), we have lost the muscles through which we otherwise could squat.[17] We do not then sit "in" chairs; our bodies are themselves, in a sense, chaired. The chair has snatched our musculature and now made us dependent on it. So chairs, we might say in the language of social theorists such as Bruno Latour and Alfred Gell, are "actants" in the world. These things have force, not only in shaping our lives but also in determining what we are. The sitting posture happens to be unhealthy for anyone; the chair-toilet interferes with healthy elimination, a significant cause of hemorrhoids and other proctologic ailments. And it can be dirty.

The restroom is full of appliances that shape our repertoires and personal choreographies. Automatic flushers replace the human will of when to flush, with what force, and how many times. Even ordinary faucets limit the volume (and trajectory) of the flow; the automatic ones structure how long it will last and what will be its temperature. Soap and towel dispensers regulate portion size. Building codes determine the types of appliance (urinal versus toilet), their height, and the width of stalls. Other techniques, such as the presence of "reminder" signs, tell employees (but usually not others) to wash their hands (in the United States often in Spanish as well as in English—a nod to bilingualism even in "English-only" environments). Public officials inspect to make sure these elements and others are in operation.

The stall materials and design dictate their own range of behaviors and, as objects often do, establish their own set of quiet mystiques. In the United States, almost regardless of building budget or function, stalls consist of hollow metal panels, weak and crappy. The panel designers do not even provide a simple flange that would allow them to overlap. This leaves open vertical seams where walls do not quite join and where fixed panels sort of meet doors. You can peek in and out. There are large openings above as well as below. The design is a tense compromise between provisions of privacy and surveillance, leaving it to each individual to negotiate his or her own settlement. Some people won't go do it at all; others, number one but not two; some do look in (but not over or under); others look but only out; and so on. The very flimsiness typical of restroom structures induces behaviors to compensate and to "fill in" socially for what is physically lacking. In the more wide-open spaces of sinks and urinals (where construction is also more solid), people still carefully monitor glances, gestures, and speech acts, enacting separations otherwise not present. Perhaps the most intense of restroom settings are the urinals that line men up with their penises exposed and nothing to do with their eyes. In such settings, "privacy is pretense," as Rose George puts it.[18]

But sometimes the restroom performs in a socially concentrated way, with a whole repertoire of sexuality coming to the fore. As the sociologist Laud Humphries described so minutely in his pioneering study, *Tearoom Trade*, of men having restroom sex, minutiae of signals, placements, lookouts, and excitements turn on the precise configuration of urinals, stalls, windows, and soundings.[19] Men do not simply have sex "in" restrooms; the facility and the erotic acts are intrinsic to one another. Like honeymoon beds and candle-lit dinners, the fixtures shape the particulars of actions, social arrangements, and participants' sentiments toward the scene and one another. At least such was the case in the homophobic '60s, in which Humphries made his observations.

Certain restrooms long defined nodes of gay life in cities, causing people to take routes and visit sites that they would otherwise not frequent, and certainly not at the times of day and night that they did. In many more ways, the distribution of public restrooms and their specific qualities affect the urban space: who goes where, when, and using what means of conveyance. Women are more likely to have a complex daily round, involving not only the workplace but also trips for the care of children and parents and for goods provision. Marking off the city with appropriate nodes of comfort stations thus has gender-specific benefit. People with disabilities, and recall the extensive meaning of that category, can take on greater risks of going out if there are reliably appropriate places to go. If authorities place restrooms underground, as they often did in England to assuage moral anxiety,[20] it closes them to the disabled. Twenty-five million Americans are incontinent (the great majority women);[21] they may choose to stay home and miss work, school, caregiving, or care receiving. Parents are more likely to give their children freedom to roam if they think there are safe and secure facilities, and parents are more likely themselves to travel with infants if baby-changing stations are known to be available. Hours of operation, whether of public restrooms or those of restaurants and shops, further influence just who can be out when and the means by which they travel. How will lactation occur?

Putting decent public toilets at bus and subway stations enhances use of public transportation rather than the private car. Restrooms at consistently reliable distances encourage walking. They support bicyclists by providing places to clean up as well as to relieve themselves. The toilets, or lack of them, hence perform the city as much as the other way around. We see at this larger scale the physical and social going beyond "interface" or "connectivity"—they are a seamless entity.

These ideas about gender, governance, and social-physical interaction—which I have just introduced without proper attribution—in fact derive from the authors in this book, scholars who come together because they believe the restroom has things to teach. The fruits of their labors are in the chapters that follow, along with the editors' selection of "Rest Stops" between them—images and brief texts that exhibit the points being made in a sharp, sometimes exuberant way. The editors have grouped the chapters as they cohere on three main themes. In part 1, authors focus on *use* of the public restroom, including "what it's like" to be there. For the Australian cultural studies scholar Ruth Barcan (chapter 2), public toilets operate as—in the words she uses to title her piece—"technologies of separation and concealment." Starting from anthropologist Mary Douglas's conception of dirt as "an offence against order," Barcan explores restrooms as spaces designed to separate people into categories as well as to eliminate, disavow, or conceal things or persons that threaten these categories. She examines how public toilets aim to keep at bay not only literal contamination but also the cultural contagion for which literal dirt so often serves as a stand-in. She considers how the senses work to determine, given people's fears and anxieties, if a space is safe or dangerous, clean or unclean. Could some of the negativity, she asks, be disarmed with a socially progressive architecture that would also help people feel safe and clean?

So Barcan invites us, in effect, to consider how reform might be possible. When it comes to the restroom, with whom dare I share? Exactly how much of it and by what physical means? By flimsy partitions or no partitions at all or sturdy walls that fully enclose? With no one watching us, will we be secure? What things will we do together (wash, primp, pee) and which not? Will we switch off by time of day? Who gets it when? How might it work given the culture as it now is or the culture as it might come to be?

One source of clues comes from comparison with other times and places. Zena Kamash, Oxford University archaeologist, unearths some of the patterns in the ancient world. As is evident by the question contained in her chapter title, "Which Way to Look?" the Romans shared many of the same concerns as Barcan's contemporaries. Building on what we do know from the digs, we can see similarities and differences in the physical conditions and their social management. People took shits together, sometimes arrayed in large numbers lining the walls of a room—but with what appears to be, it should not surprise us, careful consideration of what to reveal and just how. The rules seem to have been different for what women and men could show (and observe), but on certain important issues, such as gender segregation, the archaeological record is uninformative. We do know, from the archaeological record, that the zest for graffiti (and its private revelation) has had a continuous past. Found in a Roman toilet in a recent excavation were the Latin words "*cacavi sed culu(m) non estergavi,*" which mean "I shat but did not wipe my ass."[22] Toilet mischief goes back a long way.

In the closing chapter of part 1, Irus Braverman of the University at Buffalo Law School, reports on "toilet inspection" and related issues of public restroom surveillance and how they enter into the users' world. Toilet rules, formal ones in the law and informal ones as well, happen through inspection. Braverman asks, in a most broad way, how it is done—by city officials who check up on safety and sanitary codes and by the equipment itself (such as the automatic flusher), which "inspects" what people have done and then arranges, all on its own, corrective action. Again consistent with the thinking of Mary Douglas, whose work inflects so many of the essays in this collection, Braverman regards the combined regime of law and appliance as "technologies of purity." Users fit into this set of structures, adapting, resisting, and mastering as they go along.

Part 2 turns on access: who gets to go and under what conditions? There are specific occupational groups, taxi drivers most particularly—as taken up in chapter 5 by NYU sociologist Laura Norén—whose rounds of life turn on exactly where they can access a facility. As Norén points out, New York's drivers need a place to safely park and run in to do their business. Norén contrasts the life of cab drivers (and some others who operate in the streets) with the excretory freedom of New York's dogs. Authorities and bystanders exercise forbearance for the deposits dogs make on the sidewalk while having zero tolerance for humans in street-based occupations relieving themselves on sidewalks or alleyways. Pinched between prohibitions and missing public provisions, male taxi drivers often repurpose empty bottles as urinals. Women cannot do this, at least not as easily. The lack of accessible toilets in New York is an urban infrastructure reality that may quietly reinforce occupational gender segregation. Taxi driving is a predominantly male occupation everywhere in the United

States, but much more so in New York, as Norén points out.[23] So we have a possible explanation of occupational gender difference as well as variation in occupational structure across urban areas, rooted—at least in part—in the problem of restroom access.

This kind of blocked opportunity for women happens across a wide swath of activity, occupational and otherwise. Clara Greed writes (in chapter 6) as a professor of inclusive urban planning at the University of the West of England Department of Architecture and Planning. She has devoted much of her professional life to investigating and advocating for public facilities—especially for women. (She is also known—something she sometimes proudly proclaims—as "Dr. Toilet.") What would it take, she asks in her chapter title, to create "a nonsexist restroom?" Although there has been a sexual revolution, there has not been a commensurate "urination revolution"; for Greed, the queue for the ladies' toilet tells the wider tale. Lack of provision affects women's use of the city. In ways Greed points out, the intersection of "the bladder's leash" with urban scale and intricacy sculpts women's spatial and temporal paths. Greed calls for a "toilet strategy" that takes into account the needs of different sorts of people in how they use the city. Competent and righteous urban planning means attending to the routines of women (and others), of which toilet use is a crucial aspect.

The places that women do have emerge from particular histories of what was considered their proper role and, as taken up by legal scholar Terry Kogan in chapter 7, the growing needs of industry to have women in the workplace. Female labor was needed, and their bodies somehow had to be accommodated without defiling either the workplace or notions of womanhood still prevalent in the earlier periods of U.S. industrialization. So women were given special rooms, with particular features—in at least idealized versions, "honoring" their presumed sentiments and genteel ways. Laws requiring that public restrooms be separated by sex are thus not a simple recognition of natural anatomical differences. In Kogan's view, the first laws adopted in the 1880s requiring separate workplace toilets by sex were a last-ditch effort by Victorian regulators to recapture the moribund "separate spheres" ideology of the early century. It was an ideology that considered a woman's place to be in tending the hearth. If the realities of the late nineteenth century made it impracticable to force women to remain at home, then the law would mandate special spaces in the dangerous public realm, set aside to serve as protective havens—surrogate "homes away from home." However terrible the labor conditions for factory women, there was a modicum of chivalry—a mixed blessing, to be sure. Rather than ushering in a radical break with prior gender norms, those conventions were more or less accommodated across the work and gender landscape.

Grappling with another face of convention and access, David Serlin of the Department of Communication at the University of California, San Diego, learns from the disabled and their restroom experience. There is general hostility to bodily difference, disability included. Serlin focuses on the special stigma faced by disabled men (as opposed to women), for whom disability carries a double vulnerability. Given the common association of dependence with femininity, men with disabilities thus have a gender role to clear as they accomplish elimination and grooming. There are inhibitions to talk among strangers, especially in the men's room. It is not easy for one man to reach out, much less touch the other—a pattern signaled by Barcan in her chapter. It means that men lose potential access to helpful hands and, especially relevant for the blind, information on the spot. Men are less disposed than women to helping in the first place, or at least that is a common enough perception to guide behaviors of both potential helper and help recipient. This puts a premium on putting the help into the hardware, such as bars for self-hoisting onto the seat, and the kind of quality maintenance that makes certain that the right equipment and products are reliably at hand. It also means that the set-up must, in lieu of human support, work well for the diversity of types of disability (wheelchairs, crutches, blindness, frailty)—a particular design challenge. Providing access for the disabled means taking all such things into consideration—how the social and physical contexts work (or fail to work) as ensembles of dignity.

At a more macro level, disability joins the public restroom to raise dilemmas of what a just outcome might be for a liberal society. Gaining any kind of access has been a hard-won accomplishment of the disabilities movement. But if a goal is to integrate disabled people into the larger body politic, then how is it appropriate to mark them off with special toilets and facilities? When they are the only people served by a "unisex" facility (which is often the case), aren't they uniquely degendered? And especially relevant for the anxious men, aren't they in effect neutered? Shouldn't both disabled men and women be part of what it means to be "normal," even if—maybe especially if—that represents a reinforcement of the gender binary? That would mean disabled men accessing men's rooms (and maybe *any* stall), with then also a parallel system for disabled women.

So the puzzle arises: Should disabled people demand to be part of the convention? Or should they be the leaders of a movement to combat it? Gay and lesbian people face a similar dilemma in regard to marriage. Do they join what many people consider a reactionary or at least outdated family arrangement, or do they use the numerous instances of rich and rewarding non-marriage-based gay and lesbian partnerships to point everyone toward a better set of outcomes? Should marginality, always a good vantage point for seeing the rest of the world, be mobilized to change that world or to gain access to it? Are women more equal to men if they have parallel facilities, including their "own," or does some kind of seamless blending make for the just outcome? Serlin's larger point is that ideas such as liberalism, equality, and access are gendered to their core, involving complex and differential implications for men, women, and those of other gender identities. This is an issue running through the whole book.

Part 3 of the book points toward outcomes that may emerge to grapple with these fundamental issues, especially as they involve innovations in architectural shape and form. In chapter 9, Olga Gershenson—a scholar of Judaic and Near Eastern studies—takes us through the saga of a student-led effort at her campus, the University of Massachusetts at Amherst, to create unisex bathrooms. It was a campaign that brought uproar in the campus media and a prolonged dispute with the university administration. Opponents expressed concerns with morality, safety, and financial cost to the university in rearranging something as expensive as plumbing. Echoing earlier battles over bathroom access, these concerns reveal—even in a progressive context such as a Massachusetts university—deeply seated anxieties linked with gender and sexuality. For Gershenson, the implications go well beyond the rights of transgender people, in whose name much of the campaign was waged, to larger issues of fear on the one hand and equity on the other. It prompts serious consideration, at least as a thought experiment, of what it would mean, drawing on the call of philosopher Richard Wasserstrom, to eradicate gender altogether as basis for provision—and to build and retrofit, accordingly, both physically and culturally.

In chapter 10, Mary Anne Case, a professor of law at the University of Chicago, weighs the pros and cons of moving toward such a unisex future, including the relevant legal statutes as they have evolved. She recalls how enemies of the Equal Rights Amendment used the prospect of mandatory unisex facilities to oppose the amendment—and used it with some effectiveness, maybe because many people, including those otherwise supportive of equality agendas, highly value gender segregation at the restroom. What would, indeed, be gained or lost for women (as well as for men) if we just went unisex? Does women's having a place of their own sustain benefits of sisterhood for them? And perhaps the physical needs of women (such as a place to put that tampon) are better provided in a segregated set-up. Addressing such issues forces us to ask just what inequality means "on the ground." How exactly is it that separate but equal is inherently unequal? Or does treating us as all the same perpetuate inequality? Should the privileges of difference, such as they may be, be dispensed with? However ambivalent some people may feel, many—including feminists—still appreciate the homogeneity. It makes possible schemes, maneuvers, and delights that male dominance might otherwise swamp. Case ends by making up her mind, sharing the route of reasoning that helped her do so.

Barbara Penner, a true pioneer in toilet studies, in chapter 11 takes on the subject of toilets as a subject—and why there has been so little of it. Parallel to the zeal for purity that goes on in toilet practices, however ill-conceived and self-defeating, is the ongoing anxiety of contaminating scholarly endeavor with intellectual discourse on bodily elimination and the artifacts through which it operates. The taboo has particular impacts on architecture, high and low, with an avoidance of the issues that directly reinstates old practices—something of particular salience to Penner, a lecturer in architectural history and theory at London's Bartlett School of Architecture. Penner argues that it is not only what practitioners talk about that constitutes their disciplines but equally what is forbidden as topic. With regard to public toilets, users—whether architectural practitioners or laypeople—have an intimate relation to the artifact in a way they do not have, for example, with building elevators. But mum's the word.

As discourse strategy for her chapter, Barbara Penner offers up two provocations whose narrative effectiveness relies on the toilet silences. She brings up Alexander Kira's 1966 study *The Bathroom,* a unique—almost startling—analysis of exactly who uses the bathroom, including the public restroom, and precisely how: the trajectory of urine, for example, as it leaves the male versus female body. People in his time wanted to know if Kira was "for real," which—as a professor of architecture at Cornell—he certainly was. Kira, as Penner indicates, had to resort to ergonomic and clinical descriptions. No "shit" or "pee" defiles his text. Penner sees in his technoscientific prose the need to avoid ridicule and possible career damage. His circumlocutions display the necessity of distance from the body's dark needs, impoverishing architectural theory and architecture itself, in Penner's view. Even at this late date, Kira stands alone—and his book is out of print.

The secretness of the bathroom facility can be a way to shock or at least arrest attention—something so famously accomplished by Duchamp's urinal on a gallery pedestal. The toilet can be made to work for moviemakers as well as a sculptor such as Duchamp, and thus Penner also makes much of a particular use of the bathroom by Peter Greenaway in his 1985 movie *26 Bathrooms.* The movie has all sorts of bathrooms "in use," but with some of the situations concocted precisely because of their unlikelihood: home bathrooms done up with living-room-like furnishings and activities. But there is an "expert" in the film who, in Greenaway's mockery, maintains the separateness of the toilet from other of life's activities and interactions; he is akin to the conventional academic who studiously avoids the underpinnings of elimination. Penner's programmatic message, drawn from the Greenaway film, is to get real and take the practices of life into architectural and cultural analysis.

Finally, in chapter 12, I try to draw it all together to point out, in both intellectual and practical terms, what it would mean to think, live, and design in a more straightforward way. As my vehicle, I draw on the frustrated effort of my colleagues at NYU to create a unisex toilet facility in new quarters the university was building for our occupancy. This leads Laura Norén and myself to sketch out, literally as floor plan, what an excellent public restroom would be like, as opposed to the version that eventually came to be realized.

The overall goal of the summary, as with each chapter, is to make good on the repeated claim that the loo teaches. It shows, in regard to a crucial zone of life, how person and artifact combine in the world. Individuals approach objects with both emotional affect as well as the need for utility. They run into large-scale problems of exclusion based on class, race, or gender. The micro and the macro, as always, are in mutual (and in historical as well as real-time) determination.

The loo also teaches about silence. Avoidance, because of distaste or whatever reason, has special consequences on the nature and distribution of suffering, the givens of the physical world, and the pollution of the earth. The public restroom shows the insidious consequence of bracketing off certain truths as unmentionables. Emperors of all kinds, including grand systems of finance, war, and industrial production, have no clothes. Failure to call it out, make it visible, and discuss the implications perpetuates disadvantage, danger, and maybe even catastrophe.

Rest Stop

Russell Sage Foundation

THE RUSSELL SAGE Foundation's conference room and library—as much else in its Philip Johnson–designed headquarters building—exudes dignity and significance. Its once male, now unisex toilets appear, in contrast, as cheap afterthought.

Russell Sage Foundation library (above) and restroom (left), New York, 2009.

PART I

Living in the Loo

REST STOP
18 PAGES
AHEAD

2

Dirty Spaces

Separation, Concealment, and Shame in the Public Toilet

Ruth Barcan

I START FROM the general proposition that architecture is "an art which directly engages the body,"[1] an art that must deal in concrete, literally, with often abstract or hidden social and cultural logics. My second starting point is the well-known work of the cultural anthropologist Mary Douglas on dirt. Douglas engages with dirt as an extensive metaphor for anything that is symbolically polluting because it threatens established sociocultural categories, such as the division between male and female, human and animal, public and private. Dirt is an "offense against order," against the categories that help promote social stability.[2] It is, therefore, that which a society feels it needs to eliminate, conceal, or purify in order to preserve order. Sounds, smells, sights, objects, or even people that cross boundaries threaten the purity of social categories and are causes of psychological and social unease. The body has a special symbolic role in this social ordering, since it is "a source of symbols for other complex structures"[3] and "a reliably constant source of pollution."[4] Bodily waste is an obvious and potent form of dirt thus understood, since its potential for contamination conjoins the literal and the symbolic. This makes even the cleanest of public toilets, culturally speaking, a "dirty space." Culturally dirty spaces such as toilets produce a host of practical problems and dilemmas for architects, designers, planners, and regulators, since it is hard, if not impossible, to design a single site that can accommodate their ambiguity. There is probably no single space that will meet the often conflicting needs of different social groups, since there is no easy conformity either in principle or in concrete about what will make all people feel comfortable and safe. For this reason, public toilets are inevitably contested spaces.

So public toilets are not only ambiguous spaces and contested spaces, all the more so they are also *multiple* spaces, in the sense that they house many needs and practices. They are places for excretion and defecation, for restoring one's social face (doing one's hair or makeup), or for changing one's clothes. They are places for washing and also, for some, quasi-medical spaces (changing a colostomy bag or injecting insulin or taking medication).[5] The homeless may use them as shelter; they may be used for masturbation, drug use, or sex. Our encounter with a public toilet is thus an encounter with a host of others, as we interact on a daily basis with people who may be quite different from us but who share at least some of our bodily needs. It is an encounter too with the ghosts and shadows of the mostly unseen users who have preceded us and who will take our place. For that reason, the objects contained within a public toilet—the toilet seat, the clock, the tap, the urinal, the mirror—are especially charged with meaning and often serve as proxies for the unknown others who use the space. Public toilets are thus places where we meet members of the public and where we interact with, and continually reproduce, an *idea* of the public.[6]

My argument is underpinned by two metaphors: the spatial metaphor of the boundary and the metaphorical cluster around dirt (contamination, contagion, pollution, and purification). The two metaphors intersect, since the function of boundaries is, precisely, to prevent the spread of dirt (literal or symbolic) by demarcating, dividing, and separating people, places, and objects whose proximity might otherwise be troubling. Like all metaphors, those of boundaries and dirt are not just abstract symbols but, to paraphrase George Lakoff and Mark Johnson's famous formulation, ideas we live by.[7] At once abstract and literal, they are instantiated and brought to life in a host of everyday attitudes, beliefs, habits, and encounters, as well as in more overt forms of social regulation such as planning codes and obscenity laws. The mundane business of doing one's business in public is thus at once habitual and deeply meaningful, an embodiment of broader and deeper cultural logics.

The logic I'm investigating here is that of hygiene, understood as simultaneously a literal and a symbolic matter. The sociologist Norbert Elias argues in his classic study *The Civilizing Process* that scientific or rational arguments, in particular those focusing on the question of "hygiene," provide a seemingly irrefutable rationale for newly emerging cultural distastes. The hygienic explanation, he says, always arises after the "undefined fears and anxieties" that it claims to explain.[8] This doesn't mean that scientific arguments about the spread of disease are unfounded; it means that the question of contamination is always far greater than a set of objective biological facts. Elias's core thesis is that the progress of modernity involved an increase in shame, repugnance, and embarrassment. Through a study of European manners books from the sixteenth century onward, Elias charts the advance of what he calls the "embarrassment threshold"[9]—a steady increase in repugnance at bodily processes from nose blowing to farting, defecating, and even eating. Today, Westerners are undoubtedly more liberal in our social and sexual mores than we were a century ago, and more knowledgeable about the body, but it is doubtful that we are more *comfortable* with our bodies than medieval people were. Indeed, a sense of delicacy[10] about the body, and a certain polite distancing from it, are precisely what we would characterize as a "civilized" attitude to it. Paradoxically, this sense of disgust and repugnance at the body grew precisely as the outer world became objectively safer. Alan Hyde notes that "offence at the unbathed other is a comparatively late development in American social history"—one linked with social stratification.[11] In the United States, bathing, once thought to be dangerous to health, gradually became more common in the nineteenth century, a rise connected to the rise of individualism and the "liberal political program of self-government."[12] The fewer outer fears we have, the more we have inner fears; the more self-regulating we become as individuals, the more the fear of transgressing social prohibitions takes the form of shame.[13] Today, moral liberalization is not matched by bodily liberation; rather, consumer culture actively fuels our disgust at any body that is not tight, taut, hairless, and odorless—in short, sealed off.[14]

The sealed-off body requires sealed-off spaces in which to unloose it-self—and yet sealed-off spaces themselves generate fears. They generate desires, too, and some desires, especially socially transgressive ones, in turn generate more fears. Just as the body can never be sealed off, so its spaces can never be culturally contained, and public toilets remain psychologically and socially ambiguous spaces. Striving ideally for comfort, or at least for convenience and neutrality, they are all too often spaces haunted by displeasure, disgust, anxiety, or fear (of violence, of contamination, of arrest, of being locked in, of finding something unpleasant or distressing). In sum, public toilets can be characterized as having a number of distinctive spatial qualities: they are multiple, contested, ambiguous spaces and spaces of heightened affect and sensory charge.

The metaphor of the boundary opens up the idea of public toilets as technologies of division and separation; the metaphor of dirt allows us to see them as technologies of concealment, elimination, or disavowal, and hence as places of veiled shame, disgust, and fear. Two further points of focus are worthy of mention, although space does not permit much elaboration. The boundary metaphor allows a focus on gender and sexuality, since they are among the most strictly policed boundaries of any society. The dirt metaphor suggests the importance of the senses, since the senses are the bodily vehicles through which we detect contamination—that is, things that make us feel afraid, ashamed, or disgusted.

I hope it is obvious that this is neither an argument against sanitation nor a romantic paean to dirt. A critique of the hygiene fetish applies—indeed, is *possible*—only in contexts where hygiene is something other than a far-off dream. Making such an implicit distinction between "reasonable hygiene standards" and a "phobic hygiene fetish" is, of course, not a neutral judgment. It is itself an example of precisely the kind of cultural work of dividing and separating described here. Wherever we draw such lines, we are invoking cultural norms of cleanliness and pollution. And it is worth noting that no matter where that line is drawn, it will never make everyone happy and will involve trade-offs between the prevention of disease and the creation of new anxieties (and perhaps, if the so-called hygiene hypothesis is correct, the creation of new immune deficiencies and allergies).[15]

Public Toilets as Technologies of Division and Separation

Architecture is a practice that can make cultural separations concrete, literally. As Joel Sanders put it, "Through the erection of partitions that divide space, architecture colludes in creating and upholding prevailing social hierarchies and distinctions."[16] This must especially be the case for those architectural features that deal with hygiene, since, as Alan Hyde remarks, "social concern with hygiene is inseparable from division of the population into high and low, and control of the lower orders."[17] While this quotation makes it evident that class or social stratification is one of the forms of division entrenched in toilet design, public toilets, unlike the home bathroom, also frequently instantiate the most literal and entrenched social division—the division of people into two unchanging sexes. This form of segregation is at once immensely naturalized and immensely policed, the most taken-for-granted social categorization and the most fiercely regulated.

Gendered segregation also has an asymmetrical impact. The seclusion and individuation of women mandated by the cubicle arrangement, and the absence of any spatial design that would provide women with a functional equivalent to the male urinal (attempts to design female urinals notwithstanding),[18] while they may protect a putatively stronger female modesty, mean that women take longer to urinate than men do and are regularly subjected to long queues. As Harvey Molotch points out, regulations requiring equal floor space for men and women in public toilets enact an abstract principle of equality that bears no relation to the actual inequality of toileting opportunity in practice.[19] Moreover, this diminished opportunity to urinate is compounded by other aspects of female biology and sociality that may increase women's need for toilet facilities that are more available and more accessible—such as their frequent role as caregivers to infants, children, and the elderly; the increased need for urination during pregnancy; and, perhaps, the little-talked-about impact of serial childbirth on urinary continence.

Two anecdotes serve to illustrate this gender asymmetry. I recently passed through the major train station in Sydney on the day of a Christmas concert for senior citizens. The place thronged with elderly people. Outside the female toilets, a long queue snaked its way through the vast main concourse, an unusual sight in that venue. No queue was visible at the men's toilets. There is nothing particularly new in that except for the fact that the queue was composed entirely of elderly women. For me, this snake of snowy-haired women seemed to speak of frailty and the difficulties of aging, made all the more poignant by its so evidently gendered quality. (Of course, had I been part of that queue I am sure that it would also have been characterized by jollity, chit-chat, and patience. But it's still *unfair*.)

The second incident occurred the day after I presented at the "Outing the Water Closet" conference in New York (the event that stimulated the creation of this book), so I was more sensitive to the gender politics of toilets than usual. I attended a concert at a famous concert venue in the city, noting as I went in that there was only one set of toilets in evidence despite the thousands of patrons. Primed by this observation, I ran to the toilets at top speed at the intermission, but the queue was already long, and the ushers announced that there was a second, unsignposted, set of toilets behind a partition that they had opened specially for the intermission. Again I ran and met only a relatively small queue—perhaps ten people waiting ahead of me, among them a pregnant woman, an elderly lady, and a mother with a child. The conversation turned, as it does, to the evident lack of toilet provision and the evident availability of toilets for men, made all the more obvious by the narrow entrance hall that required the men to push through our queue to walk straight into their room, most of them looking quite visibly embarrassed at their all-too-evident privilege. By the time I left the toilets, the queue of women stretched several hundred people long, with five minutes to go until the end of the intermission.

Incensed by the fact that this scenario is repeated hundreds of times a week throughout the city, and fired up by the conference the day before, I went up to the two guards standing outside the toilets. Very politely I pointed out the long queue and asked in a respectful tone whether they thought that as no men were at this stage anywhere in sight, and as there was only five minutes to go, would they consider standing outside the men's door and making the women feel that it was quite acceptable for them to enter the space now that all the men had finished using it. After all, I reasoned, all the women around me had raised the question of entering the men's room but had been too modest (or ashamed or compliant or whatever else) to do so. One guard looked at me and replied, with sardonic politeness, "Thank you, madam. I will give that due [long pause] consideration." Then he did nothing.

To sum up, although the gendered asymmetry of public toilet design and provision is well known—indeed, it is a frequent observation in everyday life—change is slow, because it is expensive but also, as my experience at this venue seemed to hint, because the need is perceived either as an unreasonable demand or as trivial. These two stories illustrate both the depth and the naturalization of the social division of people into two sexes. I, along with hundreds of other women, chose to wait out a need rather than rupture a taboo. The strength of this taboo is evidenced in the body; it is immensely difficult, physically, to enter toilets for the "wrong" sex, no matter how pressing the need.

Inside the restroom doors, a whole host of other cultural distinctions are given spatial form. In male public toilets in Australia, the classic spatial arrangement of open urinal and closed stalls enacts a series of other binary divisions. It separates public from private and maps this division onto the body and its functions—such as that between liquid (urine) and solid (feces) or between penis and buttocks. And as I elaborate later, the spatial arrangements stage a complex theatrical dialogue between the open and the closed, the seen and the unseen, the permissible and the forbidden. But of course men's toilets have quite complex social dimensions, as I discovered in some interviews with urinal users. Some interviewees delighted in the potential for masculine camaraderie, and others were discomforted by the awkwardly social dimensions of the urinal, especially in the workplace.[20]

Women's toilets also play host to a range of bodily functions, each with its own space. Elimination and beautification take place in different zones, dimly replaying a bodily separation between interiors (which are "cleaned" by elimination in the cubicles) and surfaces (which are enhanced from the outside, with the aid of lipstick, powder, and hairbrush). After women unmake our interiors, releasing in the cubicles that which is, in the way of all abject things, neither completely *of* us nor separate from us, we remake our exteriors. This external remaking seems never to occur before the internal, psychologically cleansing, release, nor does it ever seem to occur in the seclusion of the cubicle—at least, as far as one can tell. Public toilets thus separate private, "natural" functions from public, social ones—indeed, in so doing, they actively constitute our idea of the natural and the social. This arrangement allows, and indeed obliges, women to watch each other preparing her social face ("prepar[ing] a face to meet the faces that they meet," as T. S. Eliot would have it),[21] and thus it is implicated in the public performance and pedagogy of femininity itself.[22] (This is also now true, increasingly if selectively, for male toilets in contexts where masculine body adornment or beautification has an acknowledged role.)

Women's toilets are contradictory with regard to sociality. The cubicle arrangement is at once a welcome relief from the world of appearances and an extension of it, a precious bit of solitude, a time-wasting imposition, a protection from shame, and a reinforcer *of* modesty or shame. On the one hand, women's toilets are often a site of enforced sociality, as women join the almost inevitable queue imposed by the architectural limitations of the cubicle arrangement. And yet the cubicles both assume and promote female modesty; they protect women from experiencing shame while also potentially contributing to the source of that shame, by enforcing women's bodily separateness from each other, including along class and racial lines.[23] Public toilets are thus spaces that reflect both our shared humanness—the universal need to excrete—and the social organization of humanity into divisions, classes, and castes. Some toilets, by dint of their location, tacitly, explicitly, or structurally exclude certain groups of people (toilets in expensive hotels, for example). But others throw us together. In toilets in heavily used public areas such as airports, train stations, or shopping centers, we may get more opportunities to observe people who are "not like us." Beverley Skeggs argues that toilets "heighten the sensitivity to appearance because looking is one of the main things to do when standing in a queue."[24] They are thus one site where class distinctions are assessed and policed.

The cultural work of separation is, of course, never complete, for boundaries often prove porous or precarious. As I have argued in more detail elsewhere, public toilets are fundamentally quite *ambiguous,* and hence they are psychologically, socially, and culturally *troubling* spaces. Those who do not fit neatly into social molds or categories are often faced with practical and social predicaments. Parents with a child of the opposite sex, for example, must decide at which point their child has crossed an invisible threshold into the kind of maturity that obliges them to go it alone. In Australia, a number of highly publicized violent events in public toilets have made this quite a traumatic threshold for many parents, who are reluctant to let their child enter a public toilet alone but feel constrained by the architectural norms that mandate such a separation. Despite this fear of separation, a host of factors prevent most Australian women from embracing unisex toilets—from the rigidity of the two-sex model to shame at sexual difference, fear of violence, and even sensory disgust at men's supposedly less congenial urinary habits. Similar problems confront carers for disabled or elderly people of the opposite sex.

The architecture that reinforces binary sexual segregation also produces predicaments for the transgendered. From the point of view of transgender, architectural segregation is a mixed blessing. On the one hand, what Jacques Lacan calls "the laws of urinary segregation" increase the threat of ejection, harassment, or violence against transgendered people.[25] They also, as activist Pauline Park has pointed out,[26] add social pressure to transgendered people, since part of the process of acquiring a new gender is learning the social codes of its bathrooms, codes that put public pressure on a newly reforming body image. One solution called for by some activists is to add more single-user cubicles. In that context, perhaps individuation is a not-unreasonable solution to the problem of multiple and conflicting sources of comfort—or at least a more achievable one than a dream of a welcoming communality. Segregation, in this context, is safety. And yet other transgendered activists or theorists—those with an investment in the project of unsettling or even undoing gender—might claim that such a solution simply reinforces what is to them the primary conceptual, social, and political problem: the two-sex model itself. While the provision of single-user cubicles would solve practical problems for many transgendered people, it would not address this primary separation of people into two sexes—unlike, say, unisex (i.e., no-sex) toilets, which provide a more radical, more political, provocation to laws of urinary segregation.

What, then, might a progressive and humane architecture even *aim* for, and how would it be possible to produce spaces amenable across such social divides? It is not all bad news, however. Across the contradictory claims of open versus closed spaces, unexpected congruities of interest can unexpectedly occur. Pauline Park argues, for example, that helping the transgendered by providing more single-user cubicles also helps the disabled and families with children. Perhaps, then, one answer for designers is that wherever possible, different sorts of spaces might be made available, allowing people some degree of choice in the matter of how they perform intimate acts in public.

Public Toilets as Technologies of Concealment, Elimination, or Disavowal

As well as technologies of division and separation, public toilets can also be considered technologies of concealment, an idea I derive from Gay Hawkins's reading of sewers and drains, which, she argues, make waste "disappear" and "provide a literal and moral escape from the unacceptable."[27] The senses play a special role in this drive for concealment, given their connection to emotions such as disgust, fear, anxiety, and desire. They are the means by which people decide whether a toilet is fit for use.[28] As Charles McKinney, chief of design at the New York City Department of Parks and Recreation, put it, when it comes to deciding whether a public toilet is fit to enter, "Viscerally, we make the choice."[29]

Touch, that most immediate and visceral sense, is a source of anxiety for many people. They have elaborate bodily rituals to ward off the contamination they associate with touch: they avoid touching door handles, taps, soap, and the toilet seat itself. There is, of course, some rational basis for a fear of infection, since bacteria liberated by the flush remain airborne for at least twelve minutes and also settle on local surfaces.[30] But is a toilet door really any more contagious than a café door, or a tap any less clean than a desktop? According to the eminent microbiologist Charles Gerba, the toilet seat, whether at home or at the office, is far less germy than a whole range of objects perceived as "safe."[31] There is quite a stark mismatch between our perception of toilets as unclean and the microbiological truth that everyday objects such as bus handrails, phones, and desks are usually substantially germier. This is as true of the home as it is of public places; the home toilet came last in Gerba's list of enteric bacteria levels in eight places around the home.[32] Sites or objects we associate with cleaning up—such as sponges and kitchen sinks—are typically far more germ laden than is the toilet; indeed there are typically more fecal bacteria to be found on the hands when we prepare a meal than when we exit a toilet.[33] And there is quite some irony in the fact that the lipstick that women so dutifully apply to symbolically finish off the cleansing process carried out in a public toilet is one of the germiest items in the office.[34]

To a cultural studies person such as myself, it is tempting to see the increasing fear of contamination (which, incidentally, I witness year by year in my students, who each year make me aware that I should be afraid of the germiness of more and more everyday objects and who help make me more self-conscious about my "dirty" body than ever I was as a teenager) as an embodied metaphor for a social phenomenon—that of an escalating individualization of the self and a concomitant retreat into well-separated, bounded, and sealed bodies. And the very idea of contagion is itself contagious. After reading a number of Gerba's studies I found myself pulling back from handrails and eyeing the photocopier button with suspicion. To Norbert Elias, this process of retreat is a corporeal and psychic rendering of an initially social phenomenon: new social values produce new psychic structures and bodily responses (such as visceral experiences of disgust) and ultimately new rational, "scientific" justifications. This process, whereby the social becomes psychic, finishes with a kind of cultural amnesia, whereby we forget that things were ever different and view any breaching of the taboo with disdain and disgust, as the province of uncivilized others (such as children or "foreigners").

When I hear of people afraid to touch a tap, I think less of real germs than of the fear of the other.[35] Prohibitions on touching objects—whatever their microbiological basis, sound or otherwise—inevitably involve a fear of touching the body of another, even by proxy. An increasing array of physical objects—and especially those associated with public toilets, since these are at once public and intimate, neutral and used—have the shadow of the body in them. Surely fear of "the prewarmed seat"[36] is less a rationally grounded fear of infection than a fear of the touch of the stranger, the Other who is so like us as to share our bodily shape and our bodily needs but who is unknown to us and therefore potentially contaminating. After all, the person afraid of the toilet door or tap no doubt pats dogs, catches buses, or touches objects previously touched by his or her *family*. So it is the unseen touch of the unknown person—the stranger who preceded us—that makes us afraid, even though, as Gerba puts it, "home is where the germs are."[37] So whose needs should architects and designers serve? The needs of a humanist agenda that might fight against social atomization or those of the actual humans who use these spaces and seek comfort?

As more and more technologies develop that enable us to communicate without touch (from phones to email to phone sex to virtual surgery), it seems likely that touch will become more and more stigmatized as a vehicle for contamination, both literal and symbolic. At this point, however, I find it interesting that the everyday objects that enable this distant touching—computer keyboards, phones, earbuds—are not yet objects of popular disgust in the way that, say, the toilet seat is. Phone handsets appear to be highly germy. Gerba and his colleagues found phones to have very high bacterial levels.[38] In one study, phone handsets had 25,127 bacteria per square inch, compared to 49 on the toilet seat.[39] But we ignore the forms of communication that give us pleasure, thinking of them positively as connection and "interaction," and stigmatize other, unwanted, connections.

Contagion also literally spreads through the air, and our cultural training makes us imagine we can detect it through our sense of smell, a sense traditionally associated, at one end of the scale, with the divine and, at the other, with animality, foreignness, disgust.

Unlike in medieval days, when smells were an important part of public feasts and parades, smell has few public functions left—except that of eliminating or masking the odors of bodily life.[40] Constance Classen and her colleagues understand the modern disgust at "bad" (note the moral overtones) smells and our obsession with deodorizing to be entirely congruent with modern bureaucratic society's respect for personal boundaries and its insistence on the objective over the emotional, at least in public life.[41] Though the "hypersensuality of the contemporary marketplace"[42] is rapidly altering this rational objective ideal, smell is still threatening to the individualist, hygienic sensibility, because of its "radical interiority, its boundary-transgressing propensities and its emotional potency."[43] In the cosmologies of premodern Europe, argue Classen and colleagues, smell was believed to reveal something of the inner self:

> Through smell . . . one interacted with *interiors,* rather than with surfaces, as one did through sight. Furthermore, odours cannot readily be contained, they escape and cross boundaries, blending different entities into olfactory wholes. Such a sensory model can be seen to be opposed to our modern, linear worldview, with its emphasis on privacy, discrete divisions, and superficial interactions.[44]

Consumer culture, concerned more with glittering surfaces than interaction with interiors, uses smell primarily as a concealer, the olfactory equivalent of what Eric Michaels sardonically calls the "high high-gloss."[45] In Michaels's reading of the modern obsession with the *appearance* of cleanliness, an obsession he calls "tidiness," the concern with masking and concealing is ultimately an attempt to obscure "all traces of history, of process, of past users, of the conditions of manufacture."[46]

Public toilets invite us to consume the *signs* of cleanliness, which usually have little to do with actual cleanliness or health. The chemical air fresheners that so often permeate public toilets do not cleanse or purify the air but mask the smells we associate with dirtiness using a blend of (arguably) toxic chemicals. They replace the smells of bodily interiors with the supposedly reassuringly smells of science, sterility, neutrality. As evidence mounts of the potential of such chemicals to weaken the immune system and of antibacterial products to contribute to the development of resistant "superbugs," it's time to ask whether there is a better path to olfactory pleasure than covering smells with chemicals. Could airflows be used to dispel the odors of waste? Could they be part of an olfaction of openness and revelation rather than concealment?

Another sensory stimulus that works via overflow, seepage, and intrusion is sound, which, like smell, "is no respecter of space."[47] Like smell, bodily sound is associated with disgust (for others) and shame (at self). Who has not sat silently on the toilet seat in the workplace, waiting for a colleague to leave the room before risking the shaming sounds of defecation? Should we work against the ever-increasing embarrassment threshold and learn to be proud of our sounds, or can architects free us of the shame of the sounds of bodily process that prove that food turns to waste and waste breaks open the surfaces of our body, reminding us of the thin membranes that separate purity and corruption?[48] If we were rid of the shames of smell and sound, perhaps men and women might even consent to share the same physical space. But even if it were architecturally possible to make cubicles impermeable to noises, it does not seem culturally possible. For just as the spatial separation of men and women into different rooms aims, as Harvey Molotch reminds us,[49] to reduce male violence against women, so the free circulation of sound is part of women's defenses against that same threat of violence. Knowing that your screams can be heard outside was the first thing matter-of-factly mentioned to me by a woman when I asked some of my friends whether they thought sound was an important factor in public toilet design. I had been thinking shame; she was thinking violence—two different aspects of female embodiment.

The ability to hear is also part of the policing of sexual activity. Although there has tended to be a bias toward visuality when we think about surveillance, listening is an important component of the policing of the social.[50] In the case of male public toilets, the ability of sound to break through borders plays an important role in limiting sexual activity in a number of forms, including masturbation and homosexual sex. This realization came to me rather late in my writing of this piece, since the design of men's toilets is so clearly a carefully orchestrated *visual* technology aimed at testing and policing masculine sexuality. In particular, the open space of the urinal simultaneously promotes and tests the avowed normality of heterosexuality. A range of visual mechanisms, such as lighting, mirrors, posters, and even televisions (or, more extremely, the removal of cubicle doors, as illustrated in the "Rest Stop" case that follows this chapter)[51] aims to regulate men's behavior, but conduct is mostly policed by the power of men's own gaze (a surveillance ultimately subtended by the power of law). The semiotics of the men's room is underpinned by a fundamental paradox: men are forced to put extreme effort into the appearance of insouciance. Lee Edelman vividly describes this disavowal as "ritualistic indifference," "a vigilant nonchalance" that must seem to "greet each act of genital display."[52] Moreover, argues Edelman, this is no innocent oddity: the men's room is structured to ensure the possibility of error; that is, it's structured precisely so that one can break its unspoken law—not only that one must never look at another man's penis but also that no man must ever be seen *trying* not to look.[53] Men's toilets are visual and spatial technologies designed to harness and secure the gaze and to punish aberrant looks. They simultaneously disavow and test for the presence of a contaminating homosexuality.

Disavowal is also part and parcel of the modern attitude toward human waste. Gay Hawkins argues that our modern desire for bodily purity depends on "active *not seeing*"[54]—in the case of Australia, on the denial of the subterranean drains that take our waste into the ocean. This active not seeing begins in the very locations at which we expel waste. Perhaps, then, in the process whereby public toilets bring these two forms of desired invisibility into conjunction (the invisibility of human waste, the invisibility of nonnormative sexuality), there is the potential for one to stand in for the other, for the idea of sexual desire between men to function as a kind of social uncleanliness, to be expelled or purified. After all, as Alan Hyde puts it, "every step in the history of public health measures for the encouragement of personal hygiene, the spread of bathing and eradication of filth, is always and necessarily a form of political hegemony."[55] Sexually segregated public toilets concretize a compelling overlap between the deep ordering principle of gender/sexuality and the literal problem of the elimination of human waste. In so doing, they enable a kind of subtle cultural homology whereby those who represent a threat to the established gender/sexual (and sometimes racial) order may themselves come to be imagined as a form of cultural waste—or, if that is putting it too strongly, as a form of troubling category whose presence we would rather conceal.

In thinking about this issue, it is interesting to reflect that the processes of stratification and abjection associated with hygiene agendas not only divide people into types or classes—classically, into "two worlds, a high world of refinement and a detested lower world"[56]—but also function increasingly to *individualize* us. Hyde argues that a "belief that everyone might become inoffensive" represents a graphic "individuation of the person."[57] Where once odors might have gone unnoticed or, if not, could represent only "the forces of nature or a less differentiated mass of people," now they are tied to the individual, who is understood as able and obliged to manage his or her own body odor, through the individual tailoring of deodorizing commodities and cleansing practices.[58] Personal hygiene, firm boundaries, internal purification, self-government, and, interestingly, security are now tightly interlinked.[59] This is so, but in public toilets we often do not know or even see the previous occupant, so we are back in the territory of the unknown and of imagined *categories* or styles of person—unless we enter the cubicle as they leave, in which case we are potentially granted a corporeal trace simultaneously anonymous and intimate: the warm seat, the odor, or—worse—the remnant in the bowl. In the workplace, where we may know the person passing by us, the possibility of such glimpses into the inner bodily life of a colleague—known to us and yet not intimate to us—is excruciatingly shaming.

I'd like to imagine a world in which public toilets could be at once more pleasing to the senses, more practical, and more equal. But could it be achieved without sacrificing people's sense of comfort and their security from violence? Is there any point in thinking of public toilets as sites for a progressive architecture that might loosen rather than reaffirm divisions of humans into different "types": men/women, able-bodied/disabled, heterosexual/homosexual, child/adult? After all, divisions, separations, and disavowals are all mechanisms that make people feel safe, that keep dirt in its place. This is especially the case for gender and sexual boundaries, where fears run particularly deep. And the boundaries that make us feel safe run in different directions according to our ethnicity, age, religion, sexuality, and personality. Can we work against the logic of separation yet still have safety and comfort? Can we have a comfort that is not based subtly on fear? How do we meet particular needs—say, for example, some people's need for a quasi-sterile environment for quasi-medical procedures (such as medical injections or colostomy-bag changes) without buying too much into the hygiene fetish? Can we meet such particular needs without increasing general fears and anxieties? And can we have hygienic toilets without being too strong an ally of the embarrassment threshold, which advances ever onward in the form of a battalion of deodorizers, cosmetic procedures, antibacterial wipes, and new taboos? I'd like to hope so, before I become any more afraid of bodies—whether it be my own or that of the Other.

Harvard Science Center, Cambridge, MA, 2009.

Rest Stop

Erotics at Harvard

Bryan Reynolds

THIS WAS NOT the first time that I encountered such an act of architectural intervention. But it was the first time, because of my own pressing need to defecate, that I realized the ironic ramifications of removing the doors from the toilet stalls. The most capacious and centrally located room marked for "Men" in Harvard University's Science Center is potentially the highest trafficked facility of its kind on the campus. The Science Center is a multifunctional building containing various lecture halls and administrative offices, computer and audiovisual services, and a sizable cafeteria and library. Given the popular utilization of the Science Center, why is its chief men's room so notoriously unpopular? Simply put, most people would rather defecate in privacy, in a toilet stall with a locking door, than defecate while showcased in a doorless stall before anyone who might happen to be in the room. But the lack of privacy is not all that is signified by the conspicuous absence of the stall doors in this men's room, doors that obviously were present at some point in history, as indicated by the remaining parts to their hinges.

When I inquired about the doorless toilet stalls at the Office of the Director of the Science Center, I was told that the doors were removed to suppress the gay male sexual activity that was taking place in the toilet stalls. So to quell gay male sexual activity, Harvard is willing to expose the most conveniently situated toilet space in the Science Center and consequently make public the traditionally private act of defecating. In fact, Harvard's effort to suppress gay male sexual

43

activity is at the ready expense of the purpose for which the toilet space was originally designed. The toilets in this men's room are used so infrequently now that they might as well have been removed along with their stall doors. The architectural alteration of the toilet stalls has therefore made the atmosphere of this men's room radically different from that of other men's rooms.

In general, men's rooms have a similar decorum and ambiance. While in a men's room, most men behave in an extrahomophobic manner. They go about their operations speedily and in businesslike fashion. Rarely do they communicate or make eye contact with each other. This situation is largely due to the fact that when urinating, without the employment and protection of a locked toilet stall, the phallic symbol of male power and organ for sexual pleasure, the penis, is exposed. Consequently, the penis and its owner are made vulnerable to ridicule, based on the size and shape of the penis, and to direct physical contact between the penis and another man; inasmuch as physical contact may bring about diverse and unanticipated reactions, all of which are troublesome for the heterosexual regime, it is considered taboo. Some men, however, to escape penis exposure—their own and that of other men—urinate in the toilet stalls instead of the allocated urinals. In addition to the socially problematic implication that these men have something embarrassing to hide, there is the less overt implication that these men are purposely making it impossible for themselves to observe the penises of other men. Since the men's room is culturally constructed as a normative space with regard to sexuality, which is to say that it is constructed as heterosexual space: to be caught seeing or merely looking in the vicinity of another man's genitalia is to infringe on his private space and risk being identified as a pervert or a gay man. While in the process of urinating in a public men's room, notwithstanding the often claustrophobic proximity of the urinals, never is one man to look at, speak to, or touch another man, unless it is obvious that the men involved were familiar with each other prior to the men's room encounter; and even then, penis watching and penis touching are unacceptable and forbidden. Almost any breach of this codified etiquette is likely to result in a hostile, potentially violent, interaction. To be sure, when urinating in a public men's room, a man's ability to adequately represent conventional masculinity is threatened by almost everything associated with the accessibility of his penis.

The doorless toilet stalls in the Science Center men's room, nevertheless, disrupt common preconceptions of the men's-room environment, intensifying the awkwardness and anxiety typically experienced in men's rooms, by making an already endangered heterosexual space more dynamically homoerotic. Each of the six doorless stalls in the Science Center men's room, within which a man must sit in order to defecate into its accommodated toilet, faces a urinal. This spatial relationship encourages a quite unique and provocative voyeuristic occurrence. While one man is situated on the toilet (perhaps holding his penis so that he can urinate while defecating), he is compelled, if for no other reason, since the man urinating is the only animated object within his scope of vision, to watch the backside of the man holding his penis while urinating before him. To refrain from looking at the man urinating would require a deliberate act of avoidance. The combination of the anal stimulation achieved while defecating, the observance of the buttocks of the man urinating, and the various possibilities, tangible as well as imaginative, for the penile stimulation of both parties (such as during the process of urination), along with the homoeroticism already psychologically connected to any men's room, all make this situation particularly homoerotic.

Harvard's decision to remove the doors to the toilet stalls is ironic, then, for their removal undermines the university's purpose and the conventionally understood function of the men's room in several ways. Apart from almost eliminating the usage of the toilets, it is especially ironic that Harvard's endeavor to abolish the gay male sexual activity transpiring in the privacy of the locked toilet stalls actually magnifies the overall homoeroticism of this Science Center men's room. Conversely, the doors have not been removed from the toilet stalls of the adjacent and equally conveniently located Science Center women's room. Yet this does not indicate that Harvard does not believe or suspect that lesbian sexual activity occurs in its toilet stalls. Characteristic of patriarchy is the invisibilizing and negation of lesbianism and the relegation and production of lesbian sexual practice as male heterosexual fantasy. Accordingly, when compared to Harvard's treatment of the Science Center men's room, Harvard's decision not to remove the doors to the toilet stalls in the women's room can be interpreted as a demonstration of Harvard's complicity in the invisibilizing of lesbianism. Insofar as Harvard openly acknowledges the danger that gay male sexuality poses for the patriarchal system by forcing gay male sexual practice from the men's room, Harvard responds defensively to the greater threat of lesbianism by not explicitly addressing it at all.

3

Which Way to Look?

Exploring Latrine Use in the Roman World

Zena Kamash

WITH THE ADVENT of the Roman period in the Western world came also the spread of public toilet facilities. Although private toilets are known to have existed before this time, public latrines do not seem to have been widely used. But public latrines became a regular feature of cityscapes in the Roman world. One of the most significant features in Roman public latrines is their communality; that is, there were no cubicles or screens that provided any privacy to the users who would all be sitting together. This feature of intense sharing, together with access to water for the disposal of the waste, seems to have had a strong impact on the distribution and spread of latrine use across the Roman Empire.

This chapter looks at the evidence from Roman public latrines across the empire, incorporating architectural, art historical, and epigraphic evidence to build up a picture of where and how people used latrines during this period. I first present a general introduction to public latrines in the Roman world, looking briefly at where they were located in cities and how they were designed and functioned. I then explore what an embodied experience of Roman latrines may have been, in particular thinking about whether latrines were considered disgusting by the people who were using them. Next I examine the nuances of this general picture by looking at the regional diversity in latrine use. There is, for example, a striking difference in the uptake of latrines in Italy from that in the Near Eastern provinces. This disparity demands that we scrutinize more carefully and deeply the cultural choices that lay behind these distribution patterns. Overall, I hope in this chapter to highlight some of the deeply embedded concerns that exist in humans about their biological need to go to the toilet and to explore how these concerns have been manifested by a variety of different people living in and through different networks of social values and controls.

Introducing Roman Latrines

In large parts of the Roman Empire, latrines reached their peak in popularity in the second century AD, which coincides with the mass construction of many other types of public buildings and facilities, such as aqueducts. In the majority of Roman cities, latrines were located within or next to bathhouses, probably because bathhouses would have had a good, reliable water supply, drawn from the city aqueduct. This water was needed to wash through the drain under the seating bench and supply the gutter and also any hand basins or decorative fountains that may have been present. When not located in the vicinity of bathhouses, the latrines tended to be placed close to other big water sources such as monumental fountains (*nymphaea*). This recurrent locational pattern in busy and frequently visited parts of the city suggests that most people would have known where to find a public latrine if need be. Furthermore, the central, public locations of these facilities suggest that people would have been regular users of them. Unfortunately the architectural design of toilets and their locations do not provide any firm clues as to whether Roman latrines operated on a gender-separation basis, nor is there any written evidence to indicate this. The presence of men only in a wall painting in Ostia, Italy, may point to gender segregation, but the evidence currently available is too limited to make any definitive conclusions.

The general design of latrines across the empire is surprisingly homogeneous, and the majority have similar basic design elements. The facility often had an entrance that opened onto the street. There is evidence for doors at Pompeii and Ostia in Italy; in fact, the Via

Figure 3.1. The interior of a well-preserved latrine in Ostia, Italy, showing the keyhole-shaped apertures in the stone seating benches and the gutter running in the floor of the latrine. (Courtesy of the author)

delle Forica latrines at Ostia may even have had a revolving door. It has also been suggested at Ephesus in Turkey that the latrines may have been screened from the street by a curtain. Seating, in the form of wooden or stone benches, was arranged around three sides of the room. The surviving stone examples of these benches often have keyhole-shaped apertures on which each user would have sat. It is thought that the shape of these apertures may have permitted easier access for cleaning oneself after going to the toilet. Although most latrines could accommodate around twelve to fifteen users at a time, some were considerably larger, for example, the second-century AD latrine in Apamea, Syria, had space for about eighty to ninety users. The seating benches were placed above a large drain, which seems to have increased in depth according to the number of potential users. This drain was plastered to prevent leaking and to promote the movement of solid waste through the system. In front of the seats by the users' feet, a small gutter (about a decimeter wide) ran along the floor of the latrine. This gutter contained fresh water and was most probably used to rinse sponges on sticks (*xylesphongia*), which were the Roman equivalent of toilet paper. These are also thought to have been communal and shared between users.

The individual details of latrines varied to some extent, particularly in their decoration and other accoutrements. Some latrines were a lot more luxurious than others, with elaborately carved seats and mosaic floors, but generally the décor seems to have been quite utilitarian. There is evidence for hand basins in some but definitely not all latrines. Fountains inside some latrines may also have provided alternative hand-washing facilities, as well as improving the general ambience of the facility. Finally, the ventilation of latrines was probably quite variable. The majority of the latrines were probably roofed, but "peristyle" latrines, such as the large one in Apamea, were unroofed. Presumably the presence or not of a roof may also have been geographically and climatically determined; for example, a latrine in Britain is less likely to have been open to the elements. Evidence for windows in the roofed examples is quite limited and to a large extent is dependent on the height to which the walls have been preserved. The presence of lamps, for example, at a latrine in Caesarea may point to the necessity of artificial light due to the lack of windows, though it may also simply point to nighttime use. It is also possible that latrines may have been placed away from the prevailing wind, as noted by the excavators of a latrine at Toprak an-Narlidja;[1] it is unclear, though, whether this was just happy coincidence or not.

Experiencing a Roman Latrine

The act of going to the toilet is one that is intimately tied up with our bodies, our senses, and also our emotions. In order to fully understand latrine behavior across the Roman world, then, some account must be taken of the embodied experience of visiting and using a public, communal latrine and the value judgments that that may entail, for example, whether they were thought of as hygienic in the Roman context. Recent research on the so-called forgotten emotion of disgust has demonstrated that certain elicitors of the emotion are found cross-culturally (as also indicated in the previous chapter), which may be useful when trying to envisage and evoke the experiences of people so far removed from us in cultural space and time.[2] This work by Valerie Curtis and Adam Biran, which comprised qualitative fieldwork exploring the motivations for hygiene behavior from five studies in Africa, India, the Netherlands, the United Kingdom, and in an international airport, resulted in the following five categories of disgust elicitors:

1. Bodily excretions and body parts. Feces appeared on all the lists compiled, and vomit, sweat, spittle, blood, pus, and sexual fluids appeared frequently.
2. Decay and spoiled food.
3. Particular living creatures.
4. Certain categories of "other people," notably those who are perceived as being either in poor health, of lower social status, contaminated by contact with something disgusting (including toilets, stained sheets, and fishmongers' hands), or immoral in their behavior (especially politicians!).
5. Violations of morality or social norms.[3]

There were also a number of sensory cues that elicited disgust, especially the damp, the slimy, and the stinky. The general consensus is that many of these features come out of our evolutionary past and are designed to protect us from potential threats to health and safety, but there exist culturally prescribed ways to handle these "disgusting" biological functions. The final elicitor, the violation of morality or social norms, is particularly interesting in this context, as the implication is that disgust has social functions and is then to a certain extent also a cultural construct, as well as being part of our evolutionary makeup. It seems, then, that disgust may have been a determining factor in the spread and uptake of latrines across the Roman world.

The following sensory exploration of Roman latrines, based on the archaeological remains, suggests the strong possibility that there was plenty of potential for other people to be aware of one's animalistic behavior and to be confronted with sensory and bodily cues that may elicit disgust.

It is likely that there was some degree of visual privacy in Roman latrines. Potential external viewers would have been prevented from looking inside the latrines by the doors and other screens at the entrance to the facility. Once a person was inside the latrine, the lack of screens between seats could have afforded potential visual axes between users. The amount of lighting inside the latrine would have an impact here, and that amount, as suggested earlier, may have been variable. This variability may have been governed by whether the overriding concern was for bodily privacy, ease of movement around the space, ventilation, or other cultural tastes, such as those expressed by Junichiro Tanizaki, who lamented the rise of stark, white, bright (private) bathrooms in Japan, preferring instead the seclusion provided by the shadows.[4] Another physical factor concerning visual privacy is how much of the body would have been on view. The fact that users would have been seated, meaning that the users' bodies would have been protected by the wearing of voluminous clothing that would have draped over the body, may have resulted in little that could be seen by other viewers. What constituted nakedness, though, may have been culturally variable. For Roman Jews, for example, rules on nakedness showed a gender distinction: for women, nakedness was exposure of any part of the body and hair (if married); for men, nakedness was exposing the genitals. It may be, then, that bodily privacy and nudity was not such an issue for male Jews using a latrine but was more problematic for women, who would then have been exposed to shame, which was closely linked to nakedness in Judaism.[5] This cultural construction of what constitutes being nude in front of others seems to have been somewhat different, though, in other parts of the empire. In reference to bathhouses, for example, where nudity would have been less avoidable, the Roman writer Seneca tells us that there should be *oculorum verecundia*, the restraint and modesty of one's gaze.[6] So the voluminous clothing, the averting of one's eyes, and possibly the level of lighting should have provided both physical and moral protection against being seen by other users.

Smell would have been more difficult to control. Unroofed latrines and those which may have had windows would have been provided with some ventilation from the various smells that would have been wafting around a latrine. The lack of evidence for stench traps suggests that latrines were smelly places. Smell is well documented as having strong and complex, cross-cultural cognitive effects.[7] It has, for example, been regularly associated with theories of contagion and has often been used as moral indicator to exclude and stigmatize others.[8] In nineteenth-century Britain, for example, privies were regularly reported as "nuisances" because of the stench emanating from them.[9] This miasma was held responsible for the spread of cholera by many of the public health campaigners, including Florence Nightingale. Smell, then, may have played a strong social role in latrines, affecting people's responses to one another, albeit in ways the archaeological record does not explain.

The potential for touching unknown others in a Roman latrine would have been high. The users would have been sitting in fairly close proximity to one another, and so there may have been concerns about direct contact with other bodies, even if only through their attire. Furthermore, the design of the latrines with seats makes it conceivable that sitting in the same place as, and having indirect contact with, an unknown previous user may have been problematic for some users. The deep, dark hole over which the user sat may also have elicited concerns over the possibility of coming into contact with, for example, flies or rats. The communal sponge-on-a-stick must also be included in this context. Although it would probably have been rinsed in the gutter between uses, it would still have been shared, and thus it opened up another source of potential contact with unknown others.

A Roman latrine may have been quite a noisy environment, including sounds people made when going to the toilet, of water running in the gutter and drains and from fountains, if they were provided, and also possibly of chatter between the users. Of these, the sound most likely to elicit disgust and embarrassment would have been the sounds made by people going to the toilet. To some extent these may have been masked by the sound of running water and chitchat. The likelihood of the latter must be closely tied to how privacy was conceived and constructed. Evidence from a latrine in Ostia, Italy, does suggest that people may have been talking inside the latrines. This latrine is decorated with a wall painting, known as the Sette Sapiente, which depicts seven philosophers having a conversation about digestion and going to the toilet (fig. 3.2). In one snippet of conversation, for example, one of the philosophers says, "vissire tacite chilon docvit svbdolvs" (The cunning Chilon taught how to flatulate unnoticed). This suggests that hearing and possibly smelling farting was something of a social taboo. Unfortunately, the lower section of the wall painting has not been preserved, but we can still see the heads of another group of people sitting below the philosophers. These people are thought to be sitting in a communal latrine, and they are also having conversations about going to the toilet. Although this painting is clearly meant to be humorous, it does suggest that, in Italy at least, latrines may have been places to sit and chat for a while.

This sensual and embodied analysis of latrines suggests that there were a variety of potential disgust elicitors drawn out through negotiations with smells, sights, sounds, tactile sensations, social taboos, notions of privacy and nudity, and aesthetics. Although one's visual privacy may have been maintained through physical and moral barriers, latrine environments were often damp and likely malodorous. Sitting on the same seat opens one to contact with other people's body products—urine and excrement as well as their sweat and smells. Whatever the evolutionary origins of the capacity for disgust, specifics of time, place, and architecture no doubt shape its form and substance.

Regional Diversity in Latrine Use

The picture presented so far has been deliberately generalizing to show the extensive range of forces that may have been at play. Of course, such areas of emotion and social engagement were, and are,

Figure 3.2. The Sette Sapiente wall painting in Ostia, Italy. The large figures at the top are philosophers, whose conversation is written in Greek above their heads. Below the large figures are the heads of other people, probably sitting on a latrine, with their conversation also written in Greek above their heads. (Courtesy of the author)

culturally variable, which suggests that there may have been a high degree of regional diversity in responses to Roman latrines. In what follows, therefore, I look at several parts of the Roman Empire, which can broadly be separated into areas with high numbers of latrines that were accepted quickly (Italy and North Africa) and those where there were low numbers of latrines and a slow uptake of the latrine habit (Britain and the Near East). This analysis looks not only at the spread of latrines but also at some of the epigraphic and iconographic evidence that can be employed to expose some of the nuances of the response to communal latrines, in particular highlighting the moral adaptations of the disgust emotion and how this was tied into ideas about the virtues of cleanliness, purity, and honor.

Italy and North Africa

The high numbers of latrines found in Italy and across the North African provinces from the second century AD onward suggests that these facilities were a popular and well-received element of urban daily life. It has been implicitly accepted, then, that latrines were not problematic in these areas; that is, they were not seen as disgusting or potentially dangerous. A more complex set of social and moral negotiations seems to have been undertaken, though, if one also incorporates the evidence from imagery and epigraphy. The corpus of images and epigraphic evidence associated with latrines is small, but the majority comes from these areas. This is particularly the case for protective imagery depicting the goddess Fortuna (a personification of good fortune), snakes, pygmies, phalli, and alphabet letters.[10] A painting from Pompeii (Insula IX, 21/22), for example, depicts a person going to the toilet (which is rare in itself), who is protected not only by Fortuna standing next to him but also by the snakes which hover to his sides and over his head (fig. 3.3). The painting also features an inscription saying *"cacator cave malum"* (shitter, beware of bad things). This exhortation and the presence of Fortuna and the snakes explicitly point to the fact that going to the toilet might be dangerous.

Images such as these that feature either protective devices or personifications are also common in bathhouses across the empire.[11] I have argued elsewhere that such devices were employed as a form of hazard-precaution system.[12] Such a system, which does not require a scientific understanding of disease, seems to have been brought about by a human response to nonspecific threats from disease or pollution where the threat is from potential (and not necessarily actual) contaminants, such as the contact with other body products from sharing a toilet seat, as discussed earlier.[13] These perceived threats produce ritualistic responses that have strong cross-cultural similarities, such as having lucky or unlucky numbers, saying special prayers or incantations, or taking measures to prevent harm, in this case by the use of protective devices and imagery. The use of such devices and imagery suggests strongly that there were some underlying fears attached to

Figure 3.3. The Fortuna painting from Pompeii (Insula IX, 21/22), Italy, showing a man going to the toilet who is protected by two snakes and the goddess Fortuna. (Courtesy of Gemma Jansen)

the use of public latrines in these areas, fears that, though not strong enough to prevent the use of these facilities, did give rise to particular behaviors brought on by deep-rooted disgust and concern.

In relation to this dynamic, there seem to have been sets of social rules that governed where it was appropriate to defecate or urinate. The epigraphic evidence suggests that the major concern was with visibility, which mirrors the care that was taken to protect the toilet user from being seen, a practice that can be deduced from the architectural evidence. Inscriptions inciting the wrath of the gods for people who defecate or urinate in public have been found, for example, at the Baths of Titus in Rome and on an honorific arch at the entrance to the forum in Thigibba, North Africa.[14] As public, communal latrines must by their very nature be considered to be public places, the problem here cannot be one of defecating and urinating in public so much as defecating under the social gaze. In the latrines, then, as suggested earlier, the social gaze is deemed to have been suspended and diverted, whereas such privilege was not afforded to other public spaces such as streets and *fora*. The Mary Douglas idea, that dirt is "matter out of place," must be significant in this context.[15] If we extend this definition beyond matter to incorporate behavior as well, going to the toilet in latrines can be considered to be "in place," whereas going to the toilet in any other place where one is under the social gaze is "out of place" and thus dangerous and polluting. Furthermore, the act of defecating or urinating under the social gaze is not human behavior and does not conform to human social rules but, rather, is animalistic. The low numbers of depictions of defecating humans (and more specifically those considered to be human in the Roman world) attest to this conceptual framework. Where we do find such depictions, the people are either under protection, as in the case of Fortuna offering protection and words of caution to the shitter above, or incite the wrath of the gods, as shown in a relief from Aquileia, northern Italy, where a *cacator* is being blasted by Jupiter's thunderbolt. In contrast, when pygmies are shown going to the toilet, a depiction that is more frequent than with other people, the stigma is not acknowledged because pygmies were not considered to be human and were therefore expected to be unaware of human social rules. They also had an additional role of provoking laughter, which was deemed in the Roman world to be yet another method to gain protection against evil spirits and potential dangers.

The evidence from Italy and North Africa, then, brings out two, interconnected aspects of toilet behavior in these areas. In spite of the apparent popularity of latrines, there was evidently still an underlying worry and concern about fecal material, against which one required protection. The need to control the potential pollution (both actual and behavioral) can also be seen in the construction of the latrines themselves, that is, in the designation of a particular place for these animalistic activities to be performed, where the social gaze can be temporarily diverted. These elements, which needed control and protection from the gods, were expressed religiously.

The Near East and Britain

There are several patterns which are striking in the provision of latrines in the Near East and Britain. In Britain, and to a certain extent in the other provinces of the far northern part of the empire, latrines have been found predominantly on military sites (for example, along Hadrian's Wall) or at sites with strong military associations (for example, Wroxeter).[16] In the Near East, the uptake of latrines was significantly later than in other parts of the empire, with the majority being constructed in the fourth century AD.[17] Furthermore, there are particularly low numbers of urban sites with latrines in Judaea at any point during the Roman and late Roman periods; the only known sites are Caesarea Maritima and Scythopolis. This low frequency tends to be explained and discussed solely in terms of religious taboos over nudity and impurity that would have made using latrines problematic for religious Jews.[18] There is no written evidence to support such an explicitly religious explanation for the distribution of latrines in Syria, but the slow uptake does suggest that the inhabitants of Syria found Roman-style latrines similarly problematic and therefore that religious taboos, though probably playing a large role, may not have been the sole influence on the choice to use a latrine or not.

Examining the spread of the communal bathing habit across the Near East helps tease out an explanation. The peak of construction of Roman bathhouses was somewhat later in the Near East than in other parts of the empire, which may again be tied, at least in Judea, to religious strictures concerning nudity and graven images. In spite of this, though, bathing does seem to have become an integral part of urban daily life across the Near Eastern provinces. Where latrines were introduced, however, they significantly lagged behind and were not introduced alongside bathhouses, as one might have expected. The fact that bathhouses were more easily incorporated into urban daily life than were latrines suggests that there were broader concerns beyond nudity, specific to latrine use, that need to be explored.

The history of an area and site and its overall response to Roman colonialism seems to have played a key role in the dissemination of the latrine concept. Scythopolis and Caesarea, the two sites in Judea with latrines, are among a small number of sites in the Near Eastern provinces to display all or most of the public water-supply facilities and technologies associated with a Roman way of life. Herod, a famous Romanophile, founded Caesarea as a deliberately Roman and "Romanized" city in Judea. It is maybe not surprising, then, that this is one of the cities whose public "face" of water supply had a pronounced Roman character, eventually extending as far as the construction of public latrines. The Romanness found at Scythopolis may be related to its Greek associations. Although it had Jewish neighbors, the city and, notably, its inhabitants wished to be viewed as Greek (a common Roman aspiration), as was made clear on a second-century inscription which referred to the city as "(one of) the Greek cities of Coele Syria."[19] This non-Jewish, pagan character, in particular, probably provides the explanation for the high level of acceptance of Roman public latrines in the city, which stood in stark contrast to other towns and cities in the vicinity.

Although this neatly explains why latrines were eventually used in those cities, the reluctance to take them up any sooner still remains unresolved. After all, if these inhabitants were so pro-Roman, why did they not seize the opportunity to go to the toilet in a Roman style at the same time as they avidly took on other aspects of Roman public life? This brings us back to looking behind the cultural and religious taboos in search of the motivations behind them. I suggest that two main areas may have been problematic: the use of water for the disposal of excrement and the communality of the latrines.

There were clear Jewish religious strictures against the depositing of excrement in water. Jewish law dictated that human excrement should be buried in a field. Latrines, therefore, posed a clear problem, as they "flushed" excrement into the urban drainage system. The early (private) latrine in Herod's Second Palace seems to have presented an innovative solution to this problem. This latrine was particularly unusual in that although it had a channel underneath the probable seating area, it had no connection to the bathhouse water system despite its proximity in the building.[20] The excavators have proposed that baskets may have been placed in the channel under the seats and their contents subsequently buried. This practice of using baskets to remove excrement from latrines is attested to in two Syriac texts. The *Chronicon Ecclesiasticum* tells the story of Patriarch Athanasius I Camelarius, who went out at night to the spot that the monks would go to out of necessity (that is, to relieve themselves), and with nobody looking on, he carried off the waste in a basket on his back.[21] A similar story is also told about monks in the monastery of Qartamain.[22] Although these Christian stories do not include the requirement of burying excrement, the use of baskets in both does point to an alternative method of removing human excrement without the use of a water-based latrine structure, a method which was probably widespread across the eastern provinces. In addition, there is an emphasis in both stories on the event's taking place at night, without anybody watching. It is possible that this secrecy was to avoid being accused of striving for public approbation for undertaking an unpleasant task. It may also be related to a need to perform this task away from the social gaze.

The same problem with the disposal of excrement in water may also be true of Britain, where there seem to have been longstanding associations between water and ritual, as evidenced by the large number of prehistoric metal hoards that were deposited in rivers and springs. This association appears to continue in Britain into the Roman period, when it takes on a more explicitly religious form.[23] Although attitudes toward water and its associations with life and death appear to have been entangled in a complex and multifaceted set of negotiations in the ancient world, and there were concerns about the cleanliness of the water, for example, in bathhouses, the deliberate deposition of excrement in water seems to have been a step too far in some areas. Ethnographic evidence suggests that this may have been because the deliberate contamination of water can be viewed as the contravention of basic order.[24] This essentially evolutionary concern could then become a moral problem, expressed in terms of ritual purity and impurity, as the person who pollutes could be deemed a pollutant him- or herself.

The problematic nature of communal latrines, as discussed earlier, need not just be centered around potential encounters with nudity but may be more tied up with issues of proximity and contact with known and unknown others and the roles of the social gaze in mediating that contact. In particular, it is highly likely that the issue of shame (*pudor* or *verecundia*) would also have played a key role. In practice this meant that if a Roman showed a sense of shame, for example, by withdrawing his or her gaze from a god or a person in the manner described earlier, he or she would be augmenting that god or person.[25]

It seems possible, however, that this way of constructing shame and honor may not have been shared across the Roman Empire, for example, in the Jewish areas. The rules against graven images in Jewish religious texts suggest that the gaze may have been conceived of in a different way among Jews—as something to be avoided altogether. It is perhaps relevant, in this regard, that the only early latrine known in Judea is the *private* one in Herod's palace, described earlier. Its being private is important because the risk of bringing shame on oneself through gazing on others would have been significantly reduced. Furthermore, it meant that Herod, who was both a Romanophile and a Jew (two identities that did not sit easily together and often clashed), was still able to express his Roman sympathies and desire to be Roman, while not compromising his Jewish identity. In other words, as an individual who was acting away from the social gaze, he was able to bypass the risks and dilemmas of the public sphere, whether Jewish or Roman.

This need to be outside the gaze also seems to have been present in the Christian, Syriac stories. This cultural difference in the construction of the social gaze seems to have been a quite widespread phenomenon in the East, extending beyond the scope of latrines. Houses in the Near East, for example, were introverted to the extent that the entrances were situated to prevent those inside being subjected to the public gaze. So whereas Roman houses in Italy frequently used lavish, decorative water features to draw in the gaze, Near Eastern houses (outside of Antioch) did not seem to use this technique.[26] Once again, then, the issue surrounding the communality of a latrine focuses on the awareness of oneself and others behaving animalistically, though here again there is a specific cultural filter governing what it means to be behaving thus.

In conclusion, it would appear that Roman public latrines triggered a variety of sociocultural, moral, and religious responses. The same general sets of concerns and worries appear to have been in play across the empire: fear of unknown threats of disease and a desire not to behave animalistically, particularly in the presence of others. The differences across the empire arose out of the different cultural forces that were applied to these concerns and were expressed as collective choices about appropriate forms of behavior. In areas such as Italy and North Africa protection from these concerns manifested itself in two interrelated ways. First, hazard-precaution systems against unknown and unexplained threats of disease were set in place via protective imagery and exhortations. Second, specific boundaries were drawn designating where it was appropriate to defecate and urinate or what was deemed "in place" and what was "out of place." In both cases, part or all of the protection was expressed in religious terms. In other parts of the Roman world, such as Britain and the Near East, it was not enough to call on divine protection. The potentially different constructions of where something was deemed to be a pollutant and of social rules governing the social gaze and social values, such as a sense of shame, suggests that using communal latrines presented serious problems. Where we have the evidence, these rules and concerns seem to have focused on concepts of cleanliness and purity, which were closely interrelated with moral and religious taboos on impurity; to break these taboos would have been to render oneself impure and potentially disgusting. This seems to have resulted in severely restricted latrine use in these regions.

Rest Stop

Judgmental Urinals

Interior of the Sofitel lobby bathroom, interior design by Group CDA, Queenstown, New Zealand, 2005. (Courtesy of Sofitel, Queenstown, New Zealand)

4

Potty Training

Nonhuman Inspection in Public Washrooms

Irus Braverman

Inspection functions ceaselessly. The gaze is alert everywhere.
—Michel Foucault, *Discipline and Punish*[1]

Every little thing in the washroom is monitored.
—Sanitarian, Erie County Department of Health[2]

THE INTIMACIES, PRIVACIES, and taboos of the public washroom render it almost inaccessible for direct human inspection. Especially with the decline of attendants and thus the loss of a human policeman,[3] nonhuman fixtures are set in place to do the dirty work. Moreover, in the United States, or at least in Buffalo, New York, where I have done fieldwork (and which, I think, is typical of U.S. cities in these respects), government agencies do not take up much of the slack; officials make only rare appearances on the toilet scene.

The Inspection of Public Washrooms by Humans

Responsibility for official inspection of public washrooms in Buffalo is split between various jurisdictions, depending on the nature of the establishment, including, if a restaurant, how the food is sold. Agency action is usually triggered only in response to public complaints. The former director of the Genesee County Health Department relays some of the details:

If more than 40 percent of a restaurant's sales are over-the-counter food products, then it will be inspected by the Health Department, but under 40 percent, the place is under the jurisdiction of the State Department of Agriculture and Market. This is according to an agreement decades ago between these two departments. But [the State Department of Agriculture and Market] do[es]n't have any local departments. This means that for every place that falls under their jurisdiction the level of inspection depends on their capacity at the moment. Often their inspections are infrequent and only a result of a complaint. . . . Gas stations are under the Department of Labor. I think.[4]

The Supervising Public Health Sanitarian of Erie County's Department of Health describes the infrequent inspection visits made under her jurisdiction:

Basically, the number of times we inspect a place depends on the type of food and customers. If a place is like a pizzeria, we inspect [it] once a year; if it cooks, cools, and reheats, then twice a year. We also go out on complaints. The common complaints are, there is no toilet paper or no soap. There are a few of those each month. Our secretary receives the calls, and the first thing she'll ask is, "Did you talk to the manager?" Then we'll go out to do an inspection, if it sounds serious enough. Otherwise I'll call the owner and simply tell him that there's no toilet paper.[5]

Buffalo's Associate Public Health Sanitarian depicts a somewhat more orderly and hierarchical mode of inspection:

The inspection goes through "progressive enforcement": the first level is when an inspector finds a problem and writes a report. The operator then has a certain time to correct this problem. Second level is when the inspector goes back and the problem is not corrected—then it would be referred to an office hearing, with the "district supervisor." . . . The next level is where I get involved— this is a "commissioner's stipulation." This is a formal document that says that if you want to avoid the hearing, you must correct the problem and pay somewhere between fifty to one thousand dollars in fine. Every date of inspection is another violation. A critical item is if the toilet is overflowing—then the operator must correct the problem right away.[6]

Regardless of the potential impact of such enforcements, the major regulation occurs in requirements imposed for the set-up of the space. On the federal level, the passage of the Americans with Disabilities Act (ADA) in 1990 introduced clear standards of compliance for all private businesses open to the public.[7] Other laws and regulations also govern the number, type, and form of washroom design. Although these regulations might initially seem technical and insignificant, they have moral and behavioral force.

Nonhuman Inspection of Public Washrooms

Nonhuman things are the massive proxy inspectors of washroom conduct, substituting for the direct inspection of the space, and the remediation of its problems, by humans. They also help avoid a situation in which people need to explicitly talk about and promote more stringent criteria for washroom inspection. In this sense as well, the auto-thing does the dirty work for us. It makes sure that a machine replaces the human inspection that would be difficult to sustain and to account for in this taboo-stricken place.

The study of mundane spaces and the values that go into their specific configuration illustrates how power, for good or ill, is embodied in and normalized through the design of things. This design, in turn, is reinforced and neutralized by the legal system; together the physical and legal combine as a "policing through things." Bruno Latour illustrates this idea when he describes the policing action performed by the nonhuman "features" of his car:

Early this morning, I was in a bad mood and decided to break a law and start my car without buckling my seat belt. My car usually does not want to start before I buckle the belt. It first flashes a red light "fasten your seat belt!," then an alarm sounds; it is so high pitched, so relentless, so repetitive, that I cannot stand it. After ten seconds I swear and put on the belt. This time, I stood the alarm for twenty seconds and then gave in. My mood had worsened quite a bit, but I was at peace with the law—at least with that law. I wished to break it, but I could not. Where is the morality? In me, a human driver, dominated by the mindless power of an artifact? Or in the artifact forcing me, a mindless human, to obey the law that I freely accepted when I get my driver's license? . . . Because I feel so irritated to be forced to behave well that I instruct my garage mechanics to unlink the switch and the sensor. They now invent a seat belt that politely makes way for me when I open the door and then straps me as politely but very tightly when I close the door. Now there is no escape. The only way not to have the seat belt on is to leave the door wide open, which is rather dangerous at high speed. . . . I cannot be bad anymore. I, plus the car, plus the dozens of patented engineers, plus the police are making me be moral.[8]

An aspect of such mechanical devices is that they impose the same inspection and regulatory regime on every user (at least within the bounds of particular weights and sizes). Legal standards achieve a similar sort of "due process" by making all equipment in public washrooms conform to the same specifications. These automations strive, through similarity of instrumentation, to make behavior conform across settings and social groups. Acts of standardization thus fuse two sociolegal processes that are otherwise considered separate: regulation and inspection.

A key element on the plumbing front are the codes of the American Society for Testing and Materials (ASTM), which certifies a given product or material as adhering to certain specifications of content and performance. ASTM International is a source of technical standards for materials, products, systems, and services and is one of the largest voluntary standards-development organizations in the world. It was formed in 1898, when a group of American engineers and scientists got together to address frequent rail breaks in the railroad industry. Their work led to standardization of the steel used in rail construction. Buffalo's Chief Plumbing Inspector clarifies ASTM's work in the plumbing context: "Every fixture has to have a stamp of approval. The state has a list of companies that it approves. ASTM is one of the accepted standards. If we're looking at toilets, everything has a stamp of approval right on it. We know the names already, so we don't really check the piece of paper unless something looks strange." The process is the same for other varieties of material goods. As the Chief Plumbing Inspector elaborates, "On a new piece of copper pipe there is always a name and number. Every piece of pipe needs to have the proper ASTM stamp."[9] The standard is, then, both a form of regulatory norm and an easy means for state inspection.

Another major institution that coordinates development and use of voluntary consensus standards in the United States and "represents the needs and views of U.S. stakeholders in standardization forums around the globe" is the American National Standards Institute (ANSI).[10] According to its website, the institute "oversees creation, promulgation and use of thousands of international norms and guidelines that directly impact businesses in nearly every sector." Founded in 1918, ANSI serves as "the coordinator of the U.S. voluntary standards, providing a forum for the development of policies on standards issues," and as a "watchdog for standards development and conformity assessment programs and processes."[11]

However voluntary the origins of these specifications, they become officially binding when adopted by formal legal codes, as indeed has occurred in most American states. The New York State Plumbing Code, for example, is a 144-page volume that details every aspect of what must be provided, not just in the original fabrication of an appliance but also in how it operates. This includes water use, capacities, and hygienic elements. Chapter 4 of the code, for example, asserts that the walls of water closets need to be thoroughly washed at each discharge according to ASME[12] A112.19.2M,[13] that the metal carrier supporting the bowl shall conform to ASME A112.6.1.M,[14] and that urinals shall conform to ASME A112.19.2[15] and sinks to ANSI Z124.6.[16]

In dictating the precise design of washrooms, standards indirectly control the conduct of washroom users. People learn how far from the bowl to stand when the flush occurs and how far down one must descend to find the toilet seat. The precise form and dimensions of a urinal governs how people of different heights will have to adjust their manipulations, the way men will guard their gaze, and the results for both patrons and cleaners of where the spray settles.

There are also major impacts on those with disability (see Serlin, chapter 8, in this volume). The New York Plumbing Code has a "visitability" standard to cover these issues (first published in 1961). Section 403.7 states that "signs for accessible toilet facilities shall comply with ICC [International Code Council]/ANSI [American National Standards Institute] A117.1."[17] Other requirements dictate elements such as the bar handles and making sure that every floor of a building includes both lavatory and water closet (a double-fixtured facility). But any such explicit requirements oriented toward the disabled are only the most obvious elements of the far-wider panoply of factors influencing just who can use the facility and how they will be able to do it.

Automated Flushometers, Water Faucets, and Hand Dryers

The two archetypes of the public washrooms' automated devices are the automatic toilet flusher and the electronic faucet. In "Writing Restroom Specifications?" Peter Jahrling, Director of Design Engineering for Sloan Valve, a century-old U.S.-based manufacturer of plumbing products and systems, considers "traffic, conservation, and hygiene" as major criteria for designing washroom fixtures. All three criteria seem worthy and straightforward, and they make automated features more desirable, according to the company's marketing line, at least for "heavy trafficked" washroom sites.[18]

But other reasons also underlie the choice of automated rather than manual devices, as Jahrling explains with regard to flushing technologies and his company's trade-named "flushometer":

Higher-end washroom installations must accommodate (and withstand abuse by) many users, so specifying vandal-resistant, electronic flushometers is the most effective solution. For restrooms
handling lighter traffic, manual flushometers provide the same
level of performance as their electronic cousins, but require hand-
activation, thereby increasing chances for abuse, non-flushing, and
uncleanliness.[19]

Jahrling's account links illegitimate conduct together with impurity
and unsanitary behavior, with both policed by the automated devices of the flushometer. It makes the flushing a mandatory event, at a
specified interval, ensuring that one cannot move from the toilet area
without a proper flush.

A marketing brochure produced by Sloan Valve, titled "Get with the
Program,"[20] makes the connection between criminality and toilet hygiene even more explicit. Evidently, "programmed technology plumbing," which is the name assigned to a central computerized water-usage
system,[21] has become increasingly common in prisons and is catching
on in schools as well. "A lot of people think it's 'magical mystery' electronic stuff," says the director of Sloan's programmed water technology,
"but really, it's a natural metamorphosis of the electronic faucets and
infrared sensors that are out there now."[22] "It was a natural move forward," he concludes.[23] The use of the term "natural" is perhaps odd to
describe a shift toward the artificial. But no-touch aids in the avoidance
of the bodily essence of the other, a "natural" stigmatized as filth.

For prisons, electronic plumbing abets the disciplinary regime.
"The programmed system gives the prison guards the ability to manage the use of the plumbing system," says a design engineering manager in the same brochure:

> It controls when, where, and how long prisoners get to use water
> and toilets. For instance, during a[n] inspection they don't want
> prisoners flushing contraband down the toilet; so they can shut
> down all the plumbing. Prison plumbing can also be set with timed
> lock-outs where you can flush a toilet once every 5 minutes, once
> every half hour, once every three hours, and so forth.[24]

In effect, an ultramodernist version of the Benthamian/Foucauldian panopticon is set in place.

For ordinary civilian life, microcontrol also exists—albeit aimed, according to its producers, toward being supportive of the individual user. But it assumes a standard person with more or less standard needs engaged in an anticipated standard behavior. So it is a single individual (not with a helper or child, for example) making a single bowel movement (rather than a series) or making typical movements in a stall (not preparing for an injection, for example).

This ideal user and use are programmed in. Here is a marketing description by Sloan Valve of how the equipment "knows" what the user is up to. It discriminates, for example, between a urination and a bowel movement to deliver the proper strength of the flush:

> If the user is present for less than one minute and leaves the sensing zone or chooses the small override button, a reduced flush initiates (1.1 gpf/4.2 Lpf) eliminating liquid and paper waste, saving 1/2 gallon of water. If the user is present for greater than one minute and leaves the zone or chooses the large override button, the full flush initiates (1.6 gpf/6.0 Lpf) eliminating solid waste and paper.[25]

A call to Sloan Valve's Field Service Engineer unravels further mysteries of autoflush toilet operation. He explains that flushometers use passive infrared technology, or 940-nanometer signals, which "bounce off of the (human) target."[26] Additionally, he mentions that "the infrared detector picks up the reflection of the colors of the clothes." It detects almost any color, he says, except for certain black Levi's jeans, which "for some reason we have a problem with." The toilet flushes after the "target"—which has been detected for a period over ten seconds—is no longer present. Why ten seconds? "To take into account casual traffic, say a hand reaching for toilet paper," the Field Service Engineer replies.[27] Infrared flushing technology usually works on battery power, he further explains, noting that "the flushometer's battery life provides 4,000 flushes per month for three years, or 144,000 flushes in total; and that's the absolute minimum." Under this battery-operated apparatus, the regime is never off-duty, at least in theory. By rendering human conduct in the space of the washroom the subject of clinical observation, the "auto-gaze" thus provides the expert with the relevant technical knowledge. In the process, the intimate bodily practices are excluded from those who practice them, objectified, and reduced to taxonomies and numbers.

Sloan Valve's Field Service Engineer is aware of some user distinctions that affect the equipment and its proper functioning, including by gender. He notes that women are not that fond of the automated flush system and prefer to override it with the manual courtesy flush, which they usually operate with their feet. This practice of kicking the toilet flusher, he continues, has not been very healthy for the long-term survival of the facility. Hence, in Chicago, an optical override was placed thirty-six to forty inches high, so that "only Kung-Fu masters could even think about reaching it with their foot," thereby reducing the damage incurred by foot-flush. The Metropolitan Opera of New York City took a materially opposite approach to the same "problem": instead of training women in Kung Fu, they installed floor-pedal flushers that can be elegantly operated, even in a gown and heels, with a subtle step forward. What women everywhere may have wanted, then, women in the Met received. This example emphasizes the difference between the definitions of *public* in various spaces: while the hygienic needs of the Met public are treated with respect and dignity, the more general public is a voluminous, careless, and perhaps even immoral bunch who need only be minimally accommodated.

"Ladies also like to hover," Sloan's Field Service Engineer continues to tell me in a discreet tone of voice. This practice adds another layer of complication, he says, because the infrared device reads the empty space between the thighs and the seat as unoccupied, thereby missing the presence of the "target." In order to accommodate to this practice, along with that of the standing male urinator, the infrared is now programmed to detect wavelengths at seventeen degrees down and up, which equals thirty-two to forty-two inches. In other words, it is now programmed to "see" unseated urinators, both standing and hovering.

Initially misrecognized by the auto-gaze of the flushometer, hovering women did not conform to the programmed conduct; rather, these hovering women went right on hovering. Eventually, the auto-gaze was reprogrammed to accept such conduct. In this example, then, the surveilled population has managed to alter the terms of the disciplining process and, in effect, to challenge the status of the automated washroom as an omniscient authoritarian. The nonhuman artifact, in this instance the adjusted flushometer, displays Foucault's technology of the gaze, but only imperfectly executed. Users' reaction generates a "reprogramming" that changes the object and its impact.

Cultural variation in practices need to be taken into account. My engineer informant remarks that design involves "a whole educational piece" that mandates "orient[ing] the different cultures as to how to use your facilities." The Field Service Engineer indicates that some Asian people, "unlike us Westerners," like to squat on top of the toilet seat. "We knew there was an issue when we started to find broken toilet seats," he says. Some of the problem is handled through signage. Problems of this sort also arose around the 2009 Olympic events in China: "they installed five hundred public toilets, with instructions on the wall for non-Westerners about how to use them," he says. These installations were part of a still more ambitious "toilet revolution,"[28] resulting from a three-year Beijing campaign to modernize its public toilets (at a cost of US$57 million). Currently, Beijing boasts of its 5,333 public toilets available within a five-minute walk of any downtown location. Providing education on proper use of toilets is an important task, says Ma Kangding, an official with the Beijing Municipal Utilities Administration Commission. "The good image of modern toilets will go down the pan if the users don't change their bad habits," he continues, mentioning that "some people still leave shoe prints on the toilet seats, or even take the whole roll of toilet paper away." According to Ma, Beijing dispatched eight thousand toilet maintenance staff, each responsible for a specific restroom to ensure frequent and thorough cleaning. "They also received training in hygiene standards and techniques, Olympic knowledge and practical English expressions," says Ma Kangding. The results of the selective inspections of these staff are conducted weekly, are posted on an official municipal website, and directly affect their salary.

Hand washing, linked as it is to being the antidote to the act of defecation, is an especially policed and monitored washroom activity. One is supposed to wash, especially after toilet use. In this vein, the New York State Department of Health requires establishments to post signs. In Buffalo, the signs read, "All employees must wash hands: after using the toilet; before preparing food; whenever they are soiled." Although formally directed toward "employees" only, the inscription "all employees must" is illustrated in fine print, while the words "Wash Hands" are highlighted and large. Consequently, unless one pays special attention to its finer print, the sign seems to be directed to all users.[29] The inspectors interviewed for this project mentioned that public workshops are held in various municipalities and counties in which county officials train washroom users how to wash their hands.[30]

The electronic faucet now further influences the nature of the washup. As in the case of flushometers, every aspect in the operation is centrally programmed, controlled, and examined. Delta, a large manufacturer of faucet devices, indicates on its website that "electronic faucets provide the convenience of hands-free on/off activation and help to conserve water. Electronic faucets are easy-to-use and come in a wide range of styles."[31] According to the Supervising Public Health Sanitarian of Erie County's Department of Health, "fifteen seconds is the legal time for washing your hands well,"[32] and this is the time built into the faucet. The rule of thumb, Erie's Health Sanitarian clarifies, is that in order to perform a thorough act of rinsing, one must very slowly sing "Happy Birthday."[33]

The automated fifteen-second operator excuses the user from either singing or counting. Instead, the device itself prescribes the timing. Sloan Valve's Field Service Engineer explains that automated faucets use the same principle as flushometers, only with the opposite algorithm. "Unlike the flushometer, you want it to work when the target is present," he explains. "It picks up any skin color without any exceptions," he further assures me, indirectly alluding to the flushometer's problem with detecting certain black colors. In apparent contrast to both Delta's fifteen-second and the "Happy Birthday" calibrations, Sloan appliances apparently are "set to run for a maximum of thirty seconds," the Field Service Engineer notes, "so that it doesn't [accidentally] flood a sink."[34] The downside of this system is, however, precisely its rigid fixation (at whatever timing). One must make a special effort to wash for any longer than the set period: the specific faucet will ordinarily refuse users a second auto-rinse. In this sense, the automated system also has a counterproductive effect. Instead of promoting hygiene, in line with its official designation, it may discourage effective cleanup in cases that require more than the usual flow (vomit, for example, or ministering to stained clothing). It may also foster anger toward, and sometimes even avoidance of, the facility altogether. At the very least, the automated faucet disciplines the user to speedy cleaning. This sort of mechanical operation amounts to what Foucault calls the "micro-penalty" of time, of activity, and of the body.[35] Any lateness or interruption, inactivity or incorrect gesture, confuses the system, triggering an immediate, undesired result.

Some of these features of the public restroom have infiltrated into the private household. In this vein, Delta has been promoting e-Flow, "the first hands-free electronic faucet designed specifically for residential use." Delta's marketing information suggests that "hands free, electronic faucets aren't just for commercial restrooms anymore! And Delta has an excellent selection of these wonderful faucets for home installation."[36] Delta's promotional material also indicates that electronic faucets are ideal for homes with children:

- Adjustable high-temperature limit stop lets you set the maximum water temperature, ensuring a safe level for little hands.
- Hands-free on/off activation reduces spread of germs.
- Saves water by ensuring water automatically turns off.[37]

Such advantages aside, the auto-faucet's automated and unitary mode of operation once again also excludes, for instance, those who are slower, unsophisticated, or interested in activities other than hand washing (e.g., brushing teeth, washing face). For some people and for some uses, the experience of the sink will be more alienating and challenging. As in the prison example, the enforcement of this egalitarian code on certain groups, such as children, is yet another example of a regulation of a populace that is conceived as potentially deviant and thus in need of extra governance.

Finally, I come to the last step: drying. Nonhuman machines also control when and how the act of hand drying in public washrooms may take place. Sloan's Jahrling provides the rationale:

> Drying hands completely is essential to good hygiene, and like electronic faucets, sensor-operated hand dryers encourage use while doing away with activation buttons where bacteria can collect. Automatic hand dryers are also a good source of savings in terms of operational cost versus the cost of paper towels and labor for maintaining restrooms.[38]

To make sure that the washroom user indeed makes use of the proper technology of the hand dryer, rather than resorting to the familiarity of paper towels, the hand towels are often eliminated from this space altogether. This elimination punishes the undisciplined user in advance, by creating frustration and anxiety through the absence of a familiar remedy. At this stage many users may stop resisting, instead adopting the programmed practice: drying their hands with loud auto hand dryers that seem to take forever. Others respond by not drying their hands at all, coming out with wet hands, drying their hands on their clothes, or—probably the most extreme act of resistance in this context—using the toilet paper placed in stalls. The other type of automated drying device dispenses paper towels, but one sheet at a time, in response to hand motion. Some people try to work up strategies to trick the machinery into giving up more than a single piece in a given period of time, if they can figure out how to activate it at all. And again, there is the (not so sanitary) option of wiping on clothes.

Automated Public Toilets (APTs)

The epitome in standardized washroom design is the "automated public toilet" (APT) or, especially outside the United States, "automated public convenience" (APC). Whatever the rubric, they are currently used in more than six hundred cities around the world—in Athens, Singapore, and London and, in the United States, in New York, San Francisco, San Antonio, Atlanta, Ft. Lauderdale, Florida, and more.[39] In New York City or Los Angeles, one drops in a quarter, and the door opens. According to an article in the online newspaper *Slate*, New York's new APTs have cost the city more than one hundred thousand dollars apiece; Los Angeles's cost three hundred thousand dollars, and Seattle installed five APTs at a total cost of $6.6 million (and later sold them on eBay for twelve thousand dollars as a result of reports that they were being used to facilitate illegal drug use).[40] Describing the use of New York's APT at Madison Square Park, a *New York Times* reporter likens it to a space-ship facility. It "calls to mind," the reporter says, "the sort of room one imagines adjoined the personal quarters of Capt. James T. Kirk on the Starship Enterprise. It is a 25-cent journey to the future."[41] The article then proceeds to portray one's experience upon entering this shiny facility:

> "What follows is possibly the longest and most awkward 20 to 30 seconds of a person's day. The door slips open like an elevator, but then it stays open, to accommodate those who need extra time getting in. . . . It is very difficult to look inconspicuous in a bathroom on a sidewalk in New York with the door open . . . .
>
> There seem to be as many buttons as on Captain Kirk's bridge. Red buttons, blue buttons, yellow buttons, black and green buttons. The red ones near the door and toilet call the company for help in an emergency. The yellow calls for "assistance." . . . Blue flushes. Black dispenses toilet paper. One will quickly familiarize oneself with that button, because the designers have deigned a little 16-inch strip the standard helping of paper. A word to the wise: There

is a maximum of just three helpings. Another tip: Do not tarry. A grim yellow light turns on when there are just three minutes remaining, and after that, the door will open.

. . . The big shocker here is the soap dispenser, which actually emits not a little squirt of soap, but a jet of warm water, with the soap already mixed in. Everything is motion-activated. No knobs anywhere. The warm-air hand dryer seems somewhat slow and weak, especially with that yellow light blinking by the door. Assuming one finishes before the 15 minutes are up, the big green button opens the door. The horns and sirens and chatter of the city return, jarringly."[42]

The jarring effect is probably due to the fact that most washrooms, even those in private settings, have adjacent antechambers: some form of spatial arrangement that works to soften the transition between the intimacy and quietude of the washroom and the relative raucousness of the outside. This sort of gradual transitioning device—perhaps civil inattention embodied in an architectural format—increases the impression of the washroom as a private space, making the project of inspection yet more inconceivable.

Positioned as a blessing from public health perspectives, automated washrooms pose a potential threat on other fronts. For example, Madison Square's automated toilet facility uses fourteen gallons of water for each use (mostly for cleaning the facility)[43]—this at a moment when waterless urinals are being installed across the United States and much of the world. The APT has its other problems as well, including the need either to make them very large in scale to accommodate those with disability (as must be done in the United States) or to exclude some people altogether (as is often done in Europe). In addition, automatic devices are nontransparent and do not lend themselves to amateur repair. Many of us know how to raise the toilet top and fiddle the bulb to bring a recalcitrant flush, but the autoflush is a mystery. Coaxing an automatic faucet into service is similarly beyond our ken. Especially given the high-stakes world of washrooms, the potential for breakdown and misfiring can exact a price—an alienation, silently endured, in public space.

Humans and Nonhumans "Kicking Back"

Hence we find on Craigslist:

> Auto-flush toilets, I despise you. I hate the way you begin flushing as soon as I stand up. I hate the way you won't let me get in a courtesy flush should I be recovering from a night of Indian food. Most of all, I hate the way you flush so violently that you spray little droplets of water of dubious cleanliness all over the stall, forcing me to press myself against the farthest corner, pants still around my ankles, and you, like a rogue Catholic priest, spray holy sewer water on my freshly painted toenails and lovely new Nordstrom open-toed high heels.
>
> Motion-activated sinks, I loathe you. I don't like having to bend over and hold my hands in front of you like I'm making an offering at a Buddhist temple and want to make sure that everyone sees me lighting my incense. I hate how half of you are malfunctioning most of the time. I hate how it takes 30 seconds to get the water warm enough to really get your hands clean. I hate your stupid accompanying automated soap dispenser. . . .
>
> I hate all parts of you, bathroom. I hate you so much.[44]

How can one explain the heated emotions triggered by automated washrooms? Underlying these emotions, it seems, is anger at being controlled, monitored, and watched by infrared eyes, in the "private" setting of the public washroom. Similar to Latour's anger at not having control over the moral choice of complying or breaking the seatbelt rule, there is, in this case, the anger at being deprived of the moral choice of being clean—of deciding whether to flush after every use and whether to wash one's hands and to what extent. Matthew Crawford also writes of the dignity associated with being "the master of one's own stuff" as a moral achievement regarding the relationship between man and machine. Where digitization cancels the possibility of mastery over the physical stuff of everyday life, the "spirit of inquiry" that leads to such mastery risks being put out of reach. In the automated bathroom the morality of the human as master of his or her environment is at stake:

Consider the angry feeling that bubbles up in this person when, in a public bathroom, he finds himself waving his hands under the faucet, trying to elicit a few seconds of water from it in a futile rain dance of guessed-at mudras. This man would like to know: Why should there not be a *handle*? Instead he is asked to supplicate invisible powers. It's true, some people fail to turn off a manual faucet. With its blanket presumption of irresponsibility, the infrared faucet doesn't merely respond to this fact, it *installs* it, giving it the status of normalcy. There is a kind of infantilization at work, and it offends the spirited personality.[45]

As a sanitary matter, human control over flushing may help an individual gain distance from the bacteria spray released by the flush (see Barcan, chapter 2, in this volume)—versus inadvertently having one's face (or that of one's child) near the bowl at the critical moment. On ecological grounds, some people might prefer not to flush at all after urination as a way to save water. Indeed, during drought years, Californians were encouraged not to flush after every urination, and dark pee in urinals was a sign of moral worth: "If it's yellow, let it mellow. If it's brown, flush it down" was the mantra.[46]

The quotation from the bathroom hater also brings up other questions of function and efficiency: Do people, as implied earlier, resentfully forgo washing their hands altogether? On whatever grounds, are people so intent on behaving a certain way that they develop strategies to outwit the auto-gazers? Fairly common examples of such "work-arounds" would be people who use the sink for two cycles (complete ones, rather than just the period they actually need), leaving the sink before their time is up. There are also people who, while still seated on the toilet seat, reach around (it takes some dexterity) and use the flush's manual override button. But maybe on net, people, at least some of them, like automation. After all, the pollution taboo produces a great deal of anxiety about physically encountering the dirt of the other, which can be alleviated by the automated fixtures' at least intended achievement of minimizing that contact.

In an interview, Jerome Barth, Director of Operations in New York City's Bryant Park Corporation (BPC), provides a partial answer to

Figure 4.1. Bryant Park women's public restroom stall with electronic seat-cover protector and flushometer, New York, 2009.

some of these questions.[47] "What I say is only relevant to our limited experience," he warns me in advance, then proceeds to discuss the problematic operation of the two automated public toilets that the BPC installed in New York City's Herald and Greeley parks in 2001. After seven years of operation, the BPC and the suppliers of these particular APTs (an advertising company) "had an amicable parting of ways," he says. Instead, at Bryant Park, the BPC operates manual public toilets (fig. 4.1): three stalls in the women's room and two stalls and three urinals in the men's. According to Barth's statistics, the public washroom in Bryant Park has between six hundred thousand and seven hundred thousand users per year. By contrast, the number of users in Herald and Greeley parks (not far away) amount to an average of twenty thousand per year. The difference in the number of toilets (altogether seven seats/urinals in Bryant Park and only two

of the automated kind) and in their location (Bryant Park has many
more visitors per year) cannot explain the dramatic difference in the
number of users between these facilities. What, then, is the reason for
this difference? "The biggest issue," says Barth, "is that the public just
didn't enjoy the APTs and much preferred attendant toilets [i.e., the
kind where a human is on duty], such as that in Bryant Park." He lists
several reasons for this dislike. First, he says,

> People are just not accustomed to the automated toilets. . . . Many
> people here were puzzled. They just don't get it: they don't under-
> stand how to make the water work; they go in, and they immedi-
> ately go out. Why? I'm not sure, but I saw this behavior with my
> own eyes. I think they just hit the wrong button and find them-
> selves having to exit the washroom and then pay again to reenter,
> which gets them all worked up.

Second, Barth adds, people prefer washrooms with attendants.
"Beyond making sure that the place is clean throughout the day," he
says, "the attendants send a strong message that someone is in control, that this place is safe. It allows you to let your guard down."[48]
Finally, Barth offers that "the automated toilets are designed as functional machines, not to create an environment for real people to use.
The only reason they are constructed in the first place is for advertising companies to win large bids for outside furniture." Perhaps to
further cancel out the implications of impersonal automation, Bryant
Park restrooms feature elaborate floral displays in the entry foyer, living things that exist only because they were purchased at significant
expense and continuously maintained by human hands (see "Rest
Stop: Toilet Bloom @ Bryant Park," in this volume).

Alongside the human resistance to automated washroom facilities—in whatever degree—there is also the issue of washroom fixtures kicking back.[49] As probably everyone who conducts him- or
herself in the world of even partially automated washrooms has observed, washroom fixtures do not always perform the roles that society and engineers have assigned to them. Some electronic eyes do
not notice our desperate hands reaching out for water, or suddenly

"decide" to constantly flush, rinse, or dry, with no way of stopping their operation. As with related human behavior, the question is whether nonhuman things behave this way on purpose, so to speak. Have they declared some sort of clandestine war on the human washroom fetish? Such recalcitrant behavior would not necessarily require agency on the part of nonhuman things. Rather, it would be a component of their mechanized making, a property of their physical nature.[50] Latour refers to this ability of nonhuman things to act, albeit without agency, as "actancy."[51] Although configured by humans to act in certain ways, and in this instance to police other humans, things also take on their own programs.

Numerous complaints about the APTs focus on their malfunctions: flushometers refusing to flush or to stop flushing and sinks that are automatically too hot or that overflow. In an interview, Sloan Valve's Field Service Engineer provides additional examples of the unplanned results.[52] For one, he says, the flushometer's battery can die, which presumably means no flushing whatsoever. Then, he adds, there are the parts that touch the water, such as the diaphragm and the solenoid. Anything that touches water, the engineer explains, is susceptible to being clogged or eroded, depending on the water quality in the place, which is particularly questionable when "they take water from the street, as in the case of most public facilities and old constructions." Finally, there are also maintenance issues. The infrared lens must be cleaned with soap and water only, not with heavy chemicals. These various failures can cause "run-on" toilets (toilets that constantly flush), "short-flushing" toilets (toilets that do not provide sufficient amounts of water per flush), or toilets that do not flush at all.[53] People, in carrying out their lives, dread loss of control; that is what the scholarship on risk teaches us. This is why people who fear flying and elevators do not fear driving in cars, for example. Automated toilet facilities remove human control in a risk-averse environment.

Malfunctions of automated washroom fixtures make one acutely aware of their omnipotent power. First and foremost, this power is a reflection of the human agency that has created the object as such, storing knowledge and programming things to behave in a certain way. Such power is embodied in the physicality of the thing and thus taken for granted and rendered invisible. At the same time, there is also the power of things that "run wild." In this sense, it is the non-human that takes control over and governs the everyday conduct of humans. In the course of performing the utmost basic human needs, humans must constantly negotiate their relations with nonhuman things. This makes for a humbling, some might even say humiliating, experience.

Beyond the "usual" form of inspection, that conducted by human agents, washroom inspection also takes place through the official adoption of a web of voluntary standards set by various national and international networks. Through the design of automated fixtures, human agency is embodied into and at the same time replaced with nonhuman actancy. Instead of placing a *human* policeman to make sure that the user flushes after every use—which might constitute an illegal, immoral, and also economically impractical act in the context of the public washroom—a nonhuman thing performs the task. Automated flushing, rinsing, soaping, and drying devices—and recently also automated doors—are the authorities.

This spatially mandated public hygiene constitutes morality in practice, one that doesn't always resonate well with the public. Aversions toward being inspected seem especially strong when the inspection is carried out in the quasi-intimate space of the public washroom. Moreover, the programmed automatic fixture's assumption of the "normal" undermines human ingenuity, inventiveness, and adaptation. Since the programming of nonhuman things is meant to accommodate the "normal" user, it is bound to exclude. This form of exclusion, though seemingly an error in nonhuman programming, is in fact an essential and inevitable aspect of the automated operation.

Equipment automation induces humans to become more automated and uniform as they use the public facility. The automated infrared gaze is, as Foucault says, "to induce a state of conscious and permanent visibility that assures the automatic functioning of power."[54] However, unlike the Foucauldian gaze, the infrared auto-gaze not only makes one *feel* constantly watched; one *is*, in fact, being constantly watched. Perhaps unsurprisingly, various forms of human resistance to these impositions have mushroomed here and there:

acts of vandalism directed at automated fixtures, their routine avoidance, or strategies of finagling how they operate. Beyond official laws enacted by parliaments or government officials and applied within a relevant jurisdiction, the things themselves, however much obscured behind a set of technical specifications, also regulate consciousness and behavior.

Rest Stop

Times Square Control

Bathroom attendant's homemade numbered paddles, New York, 2009.

NEW YORK'S BUSIEST subway station underwent rehabilitation at the turn of the millennium, including the installation of four adjacent unisex stalls. One of the attendants (a retired police officer on duty Monday through Friday, 7:30 a.m. until 3:30 p.m.) took it upon himself to devise a way to direct visitors to an available stall without speaking. At his home workshop he constructed numbered rotating paddles that indicate to which toilet the next guest in line should proceed. Part of his motivation, he told us, was to ease the way, especially for foreign tourists who cannot understand verbal instructions in English.

Times Square subway station restroom: attendant's booth, New York, 2009.

Bathrooms 1–3 are tiny, barely big enough for a person, a toilet, a roll of toilet paper, and a sink to coexist. To afford privacy, the doors span from floor to ceiling. The flooring is unfinished stone, so rough that the layer of bathroom effluvia underfoot apparently cannot be scrubbed away. The combination of small size, weak lighting, lack of ventilation, and unscrubbable flooring is inhospitable. The attendants watch electronic timers that tick down with the closing of each stall door. After five minutes an alarm sounds in the booth, and the attendant can use an intercom to instruct lingering users to finish up and vacate. If they do not comply, the attendant can unlock their door remotely.

The facility was built by Boston Properties as a condition for development rights for the office tower above.

Inside the bathroom attendant's booth, Times Square station, New York, 2009.

PART II

Who Gets to Go

5

Only Dogs Are Free to Pee

New York City Cabbies' Search for Civility

Laura Norén

FOR NEW YORKERS whose work sites are unplumbed, and mobile taxi drivers in particular, having no place to go presents a daily
struggle to maintain health, dignity, and a clean criminal record. The
diminishing number of public restrooms in the city is a failure of
provisioning whose consequences are strengthened by prohibition
policies—being without a bathroom is bad enough, but what's worse
is being summoned to court and fined for resorting to public relief.
Five city departments (the Department of Sanitation, the Department of Parks and Recreation, the Metro Transit Authority, the New
York City Police Department, and the Department of Environmental
Protection) are authorized to levy fines for public urination and defecation. A telling wrinkle in state and city legal code is that canine excretion is subject to fewer restrictions than human public excretion is.
Dogs are free to pee; people risk legal sanction for the same behavior.
Is the problem the pee or the people?

Most New York City residents rely, willingly or otherwise, on their
workplaces and their homes for the majority of their bathroom needs.
Tourists face greater difficulty, being outside the daily routine of home
and work and in an unfamiliar city. Spending vacation time hunting
down a restroom is a temporary burden; workers whose workspace
is on the street search for relief every day.[1] Fruit-stand, food-cart, and
street-fair vendors, bike messengers, construction workers, newsstand
operators, and dog walkers, as well as taxi drivers, have the nowhere-
to-go problem. They hope small retailers and restaurant managers are

willing to overlook "employees only" bathroom policies (fig. 5.1); they surreptitiously use makeshift urinals out of plastic bottles and jugs. A rare, strictly emergency solution is to find a place to go on the street. Joining the free-peeing dogs risks not only legal sanction and a fine but also an uncomfortable recognition of oneself as out of order.

Recognizing oneself as the source of symbolic disorder can be liberating and empowering—social activists breaching gender binaries and fighting for unisex bathrooms may be exhilarated, proud of having stepped across an exclusionary boundary even as they fear retaliation. But New York's street-based workers have a different project; they are trying to steer clear of official sanction and make a living. Many are new immigrants working to construct themselves as regular folks and hardly in a position to readily engage in public protest on behalf of access to "rights" that are only ambiguously present in the first place. They contend with post-9/11 xenophobia, racism, and the difficulty of being near the bottom of the economic hierarchy.

The whole subject of restrooms is not voiced with ease by these workers, and this includes speaking with a white, female graduate student like me. I was able to interview fifteen New York cab drivers, who, after some reassuring gestures on my part, openly complained about having no place to go. But none confessed to the widely known practices of compensation, such as peeing in a bottle or taking some other undignified act in desperation. Many interviewees enthusiastically faulted the lack of public provisions as a direct cause for expensive parking tickets, dehydration, high rates of diabetes, poor relationships

Figure 5.1a. Bar sign: "Restrooms Are for Customers Only."

Figure 5.1b. Restaurant sign: "Restrooms Are for Patrons Only."

Figure 5.1c. Residential sign: "Do Not Urinate in This Area! People Live Here! It Is Punishable by Law!"

with hotel valets, and the occasional preference for night shifts, when parking-reliant bathroom breaks are easier, even though day shifts are more lucrative. Interviewees did speak of "friends" who had resorted to peeing in bottles or even on the street. The number of bottles full of urine near parking spots frequented by cab drivers indicates a more widespread practice than the interviewees led me to believe. An NYU student-researcher, Hillary Marcovici, found a gas-station manager whose close contact with drivers revealed a well-articulated strategy for hiding urine collected during a shift. Clear water bottles and coffee cups were not first rate, her informant told her, because empty detergent bottles have better qualities. They have wider openings, and "the smell of the detergent helps mask the urine."[2] The opacity of detergent jugs is also a positive feature, hiding urine's telltale color.

From the perspective of my interviewees, the cab driver is in a tough spot: "You get a $115 ticket if you leave your cab while it's in the queue" (as when waiting at transit-station pickup spots). "The valets, they could let you in [to the restroom], watch the cab, but they don't." "Sometimes you just think it's easier not to drink so much [points to a bottle of water in the cup holder]." My transcripts of the interviews reveal that the drivers usually used first person, but when it came to talking about the bathroom, they dropped the first person, distancing themselves from the uncomfortable predicaments being described. Notice that valets could let "you" in while watching "the" cab, instead of letting *me* in while watching *my* cab, and that *I* don't think it is easier to be thirsty than to find a place to go, *you* do.

The lack of public restrooms presents not only personal hardship for the street-based work force but also collective consequences that give specific gendered contours to the space of the city. The constraints that promote and prohibit entry into these street-based occupations, of which toilet access is one, not only are concerns for those within these occupations but also affect the collective production and experience of the city. For the occupation of taxi drivers, the lack of public restrooms helps produce an almost exclusively male occupational niche that in turn casts a gendered shadow across the cityscape. Without more than a handful of women—just 2.5 percent of hack licenses are held by women—working in the thirteen thousand taxis (or operating the three thousand vendor carts) on the streets, the streetscape skews masculine.[3] Driving a cab is an overwhelmingly masculine occupation everywhere in the United States, but nowhere more than in New York. A survey conducted in 2000 showed that women constituted 28 percent of taxi drivers outside of U.S. metro areas, where parking and finding a bathroom are easier. In fact, women's participation in the industry has been increasing since 1960, but not in densely developed cities such as New York and Chicago.[4]

One formerly available place to go, subway-system restrooms, has virtually disappeared. There were once 1,676 public facilities in the Metro Transit Authority's subway system, but now just 10 functioning public restrooms remain.[5] The parks system now maintains the largest population of public restrooms in the city—roughly 1,100 restrooms in 1,700 parks that occupy 14 percent of the real estate in the five boroughs.[6] Within the parks system, restroom buildings are often recessed from the street, folded into the site's greenery to allow visual consumption of the verdant landscape by passersby. The bathrooms, then, are too far from the cabs and their drivers to be practical—very few cabbies would be able to nip in and out quickly enough to avoid a ticket while double parked. Double parking is often the only kind of parking during the workday. Legitimate parking spots might be available at night, but by then the parks and their bathrooms have closed.

In the city's small, more utilitarian parks, restrooms are generally placed adjacent to playgrounds, efficiently colocating plumbing infrastructure for the playground's water fountains and the bathroom. This placement subtly indicates that at least when children are present the restrooms are for the kids. Signs at playgrounds in city parks prohibit adults not accompanying children from being in the playground area at all. Technically, the nearby restrooms are open to the general public, but the social stigma associated with pedophilia is so strong that these bathrooms are socially hazardous zones for all but children and their caregivers. Tucking restrooms away to preserve the visual landscape in large parks combines with the pedophilia stigma in the playground/restroom areas of the smaller parks to produce parks suitable for leisure seekers in a city left with few public restrooms that are convenient for busy workers such as cab drivers.

Compared to operators of newsstands or carts or even seasonal Christmas tree sellers, who can build relationships with neighbors to generate toilet access, cab drivers are at special disadvantage.[7] All drivers with whom I spoke revealed that each had invested time and social capital building a personal list of three or four small shops and restaurants where they could reliably use the facilities. But making friends with retailers and restaurant managers in a handful of locations around the city is not always sufficient. In Melissa Plaut's 2007 memoir of a year driving a yellow cab, she writes of being caught out of range of her regular spots, once using the bathroom at the home of one of her fares. She also describes another occasion:

> I parked at a fire hydrant, put my flashers on, locked up, and ran over to the store, just barely holding it in. Of course, they didn't have a bathroom. I tried the deli next door, but the cashier said, "No bathroom." I was doubled over, about to lose it, and I pleaded with him. "It's really an emergency. *Please.* I mean, where do *you* go to use the bathroom?" He looked at me coldly and said, "Employees only."[8]

Plaut's experience highlights a theme echoed in the interviews. The areas with good fares have almost no available parking, turning a trip to the bathroom into a game of chance with the city's parking-regulation enforcers, not to mention a potentially humiliating negotiation.

Taking Care of Business without Losing Business

> A taxi moves all day, and exchanges passengers constantly. It stops, it starts, people enter, people exit, luggage comes in, luggage goes out. Through it all, the driver remains at the wheel, like a worker tethered to his desk.
>
> —Paul Goldberger[9]

The pressure is on. In New York City, taxi medallions sell at auction for upward of $500,000 apiece, the majority going to fleet operators, not individual taxi drivers. The fleets then lease the cabs to some of the forty thousand hack-license holders, who pay about $700 weekly to drive a cab during one of the daily shifts. Working six days a week, drivers have to make about $150 per shift just to break even. On average, it takes about four to five hours of each ten- to twelve-hour shift to break even, but fares are unpredictable, and drivers are often anxious that they will not be able to recoup their weekly upfront payment to the fleet operator.[10] Each cab travels about sixty-three thousand miles annually, over half in the borough of Manhattan alone—remarkable considering the island is only about twenty miles long and two miles across.[11] A 1982 taxi ridership survey described rides as "mundane," where "the typical cab rider was a thirty-five-year-old white woman who earned $27,500 a year [1982 dollars, twice the national median income], lived in Manhattan below Ninety-sixth Street, and was very uncomfortable taking mass transit."[12] More recent data is less granular, but it echoes the earlier findings. In 2006, 60 percent of all fares were women, and "more than 85% of all taxi trips began or ended in Manhattan."[13]

In contrast to passengers' whiteness, affluence, Manhattan-centricity, and gender bias toward female, only 2.5 percent of hack licensees are women, and just 9 percent were born in the United States. Nationally, the proportion of women drivers-for-hire is five times higher than in New York and growing. New York has had the same overwhelmingly male driver population since the first survey was made in 1980. What's more, the 2.5 percent is the proportion of female hack-licensees and does not accurately reflect the proportion of women who are actually behind the wheel, which is even lower (1 percent).[14] Data on drivers' residential distribution is unavailable, but housing prices in Manhattan below Ninety-sixth Street, where most pickups and drop-offs occur, are high, out of the reach of a taxi driver's income. One driver admitted, "I had never even been to Manhattan until the day I picked up my cab for the first time. Then I had to ask directions to get to the Brooklyn Bridge from the garage."

Cabbies compete with many other searchers for parking. One analysis of traffic data found that 30 percent of the traffic on the streets of Manhattan consisted of people trying to find parking, an indication of just how much time and business could be lost by a cabbie

Figure 5.2. Water bottle, aptly branded "Nirvana," now full of urine.

trying to find a place to park for a bathroom break.[15] Both New York Police Department (NYPD) traffic enforcement officers and Taxi and Limousine Commission (TLC) enforcers are authorized to give tickets. One driver said, in the characteristic framing of what *other* drivers might do, they would "just go in a bottle, you know. Sometimes it's desperate." Physical evidence of this practice can be found across Manhattan below Ninety-sixth Street in the form of discarded bottles holding urine (see fig. 5.2). Taking care of business in the cab avoids fines for double parking, a summons for public urination, and the judgmental attention of passing locals. But it is hardly comfortable to pee into a bottle while in a quasi-sitting position. Women's anatomy and the configuration of the steering wheel makes it difficult for them to use plastic bottles or even urinals specifically designed for women (see Penner, chapter 11, in this volume), requiring female drivers to stop, find a parking spot, and then find a bathroom, all of which cuts into profits and increases driver anxiety in an already demanding work environment.

The authorities have made some modest effort to relieve the stress of cab drivers' long shifts by setting aside parking spots at a handful of locations throughout the city just for cab drivers, who can park for free for one hour. Called "relief stands" (see fig. 5.3), they do offer a rare free parking spot, but parking is the only relief on offer. They do not offer restrooms, trash cans, overhead protection from the sun or rain, a place to sit, or shelter from the elements. Neighbors are quick to complain if cab drivers litter, urinate, play music, or talk loudly. One Upper East Side neighborhood group successfully eliminated a taxi stand in its area, citing unsafe crosswalk conditions near a school, a claim taxi drivers found difficult to accept.[16] In Baltimore, an agreement between the yellow-cab organization and city hotels grants drivers access to hotel-lobby bathrooms while their cabs are parked outside, a sensible solution that has not been adopted in New York.

The leader of the successful 2007 effort to redesign the New York taxi vehicle (shifting it away from the Ford Fairlane sedan to a compact SUV-type), Deborah Marton, writes with verve that "hailing a cab—with its promise of freedom, power, and anonymity—is the quintessential New York act."[17] If experiencing freedom, power, and anonymity are quintessentially New York, then it is the rider, not the driver, who becomes an assuredly mobile New Yorker once inside a taxi. Indeed Biju Mathew, champion of the taxi workers' alliance, found that drivers experience the opposite of Marton's quintessential New York: "driving a yellow taxi in New York City makes you publicly visible and exposed every minute of your workday."[18] Mathew quotes Rizwan, a Pakistani taxi driver who likens his position to the oldest profession: "You know what the commonplace word is for prostitutes in Pakistan, don't you. . . . It's *taxi!*"[19] Without Rizwan's linguistic zest, my interviewees struggled to express the dehumanization they felt as they spent their days being directed around the city, treated as means rather than ends, a conduit from one meeting to the next, from home to office and back. One driver insisted that driving a taxi is not a "nice" job, that it is "not for good people," that it "changes you, makes you not a nice person," but that this is not because the job

Figures 5.3a and b. Taxi relief
stand: "No Standing Except
Authorized Vehicles. Taxi Relief
Stand 1 Hour Limit. No Parking
Anytime Except Taxis," New
York, 2008.

is unsafe. Another driver, only four months behind the wheel, said he likes to try to find fares going a long way, even if he knows he might not get a return fare, so that he can "talk to them, you know. Because just when you're in the city, it's so quick, you can't talk, you know. It's more relaxed to get a long fare, talk to them a little. Otherwise, it's just, you can't do the short trips, just pick up, get out, pick up, get out, you know, just too hard that way." Cab drivers often talk on mobile phones while they drive, either to family or to other drivers, sometimes asking for directions or updates on traffic conditions, other times just trying to cope with long hours of dealing with customers who may be dismissive, cheap, and incapable of providing accurate directions.[20] The law now bans such cell phone use by taxi drivers, with two-hundred-dollar tickets being issued against cabbies who talk while driving.[21] Richard Sennett and Jonathan Cobb write eloquently of the hidden injuries of social class in jobs with little autonomy as quiet assaults on personal dignity.[22] For New York's cab drivers, these hidden injuries crystallize around the prohibition of their human bodies in public space.

Peeing Legally

As a legal matter, public urination and defecation are subject to a variety of sanctions depending on the source of the excreta and the location in which it is deposited. In New York State, dogs are not, in fact, legally free to pee anywhere. Under an ordinance dating back to the turn of the nineteenth century, all animals are to be "curbed" or brought to the gutter to do their business. The law was written to ensure that the manure of work animals was kept in the street, where it could be more easily collected by the sanitation department.[23] Because these laws were commonly interpreted as a requirement to prevent the feces of work animals from befouling the sidewalks and to ensure that their carcasses would be promptly removed, the social meaning of that code never quite made the translation into a rule about pet excrement, even though the letter of the law might have been easily applied when pets replaced working animals as the city's most common four-legged occupants.[24] Regardless of the letter of the law, enforcement and public opinion permit canines to pee on sidewalks, buildings, garbage bags, fire hydrants, trees, cars, and most anything on the street.[25]

People *are* prohibited from peeing anywhere. Only designated restrooms will do. Violators are subject to apprehension and prosecution, but in varying ways depending on which government agency is involved. If a person is caught peeing in the subway by MTA employees, the punishment is only a $25 fine and ejection from the system.[26] But if caught by the NYPD, the most common urination-violation ticketing agency, offenders are subject to a fine of $250. There can also be a citation for lewd behavior or indecent exposure, which are far more serious offenses. In some states, such an offense requires registry on a list of known sex offenders for twenty years, though New York has no such requirement. A person who pees, defecates, or fails to clean up a pet's feces in a city park is subject to a fine that can be as much as $1,000, with apprehension possible by the city parks department or city police.[27] The Department of Sanitation has a Canine Task Force of fifteen assigned to apprehend non-poop-scooping pet owners and write them $250 tickets, but they cannot ticket humans for defecating in public.[28] Technically, the Department of Environmental Protection (DEP) can also levy fines for improper disposal of human and otherwise-hazardous waste that threatens public health, but this law was written to address large-scale violations, not single droppings. Additionally, the DEP does not have a street-level enforcement team similar to that of the Department of Sanitation. However, the courts are able to charge repeat excrement offenders under the DEP statute, which increases the total value of any fines at the discretion of the court.

Briefly summarizing the complex legal situation: public urination is an offense subject to a maximum fine of $1,000 and a potential charge of public exposure or lewd behavior if undertaken by humans, but there is nothing to worry about when the urinator is a dog. Defecation is more evenly prosecuted, though people may be ticketed for defecating at all, whether or not they then deposit their waste into the trash, whereas pet owners will only be ticketed for failing to clean up, not for simply allowing their pet to defecate in the view of the public. The social construction of proper bodily etiquette is revealed both in the variation between canine and human urination freedom in public space and in the potential for human street urinators to be charged with indecent exposure or civil disobedience, categories of crime punishable as serious breaches of collective etiquette.

Health

> This city is like an open sewer, you know, it's full of filth and scum. Sometimes I go out and I smell it and get headaches it's so bad.
>   —Travis Bickle, driving a New York City yellow cab
>   in Martin Scorsese's *Taxi Driver* (1976)[29]

Urine and feces present two distinct sets of concerns for public health. Urine is sterile but nitrogen rich, whereas feces is both nitrogen rich and full of pathogens. Kira notes that historically urine was valuably repurposed as "one of mankind's oldest known forms of 'soap,'" since oxidized urine produces ammonia compounds that are a standard agent for dissolving fats."[30] Because it is generally sterile, urine poses no risk to human public health. Even in cases in which the urinator has an active bladder infection, it is highly unlikely that the infecting organism can survive for long enough outside the body to come into contact with the appropriate tissue type on a new host to cause infection. The most likely source of new bladder infections is *E. coli,* an intestinal bacterium believed to migrate from the anus, mere inches away from the urethra, and bladder infection has never been reported to be caused by contact with urine on the street.[31] Untreated feces, on the other hand, carry numerous infectious agents, a real, though easily exaggerated, threat to public health. John Snow's early research on cholera in London's Broad Street and the discovery of the link between untreated sewage and new infections established sanitation as a major tool in the fight against infectious disease.[32] It is the segregation of potable water from sewage that is necessary to comply with sanitation standards. Removing refuse—even feces—from the street has much more to do with quality of life than with public health.

Figure 5.4. Sign in alleyway: "Pet Waste Transmits Disease. Leash Curb and Pick Up after Your Pet. No Dog Walking Permitted," New York, 2009.

The hard-fought battle to institute a statewide pooper-scooper law in the late 1970s was won in no small part due to inflated concerns over public health. The author of a book about the legal battle, Michael Brandow, discovered only a weak link between dog feces and sick people. "*Toxocara canis,* the common roundworm found in dogs, was known to have possible but extremely rare health effects upon children in frequent contact with their feces."[33] It was this concern that finally tipped the scales for passage of the first statewide pooper-scooper law, despite the very small number of children (two) who had contracted *Toxocara canis* from any source. The rarity of actual *Toxocara canis* infections in New York's children was less important than symbolic disgust. Current signs in the city continue to reflect the overblown link between improper pet-waste disposal and human illness (fig. 5.4).

There is an emergent secondary public health concern, if trees (and those who love them) can be considered part of the public. For the trees, urine is a much more serious threat than feces is.[34] Activist groups have recently begun to protest the practice of letting dogs pee on street trees, claiming that the concentrated urine will wilt and eventually kill the city's "lungs." Figure 5.5 shows the sign posted in a TriBeCa neighborhood, exhorting fellow dog owners to keep their pets from peeing on the trees. Evidence supports the claim that the

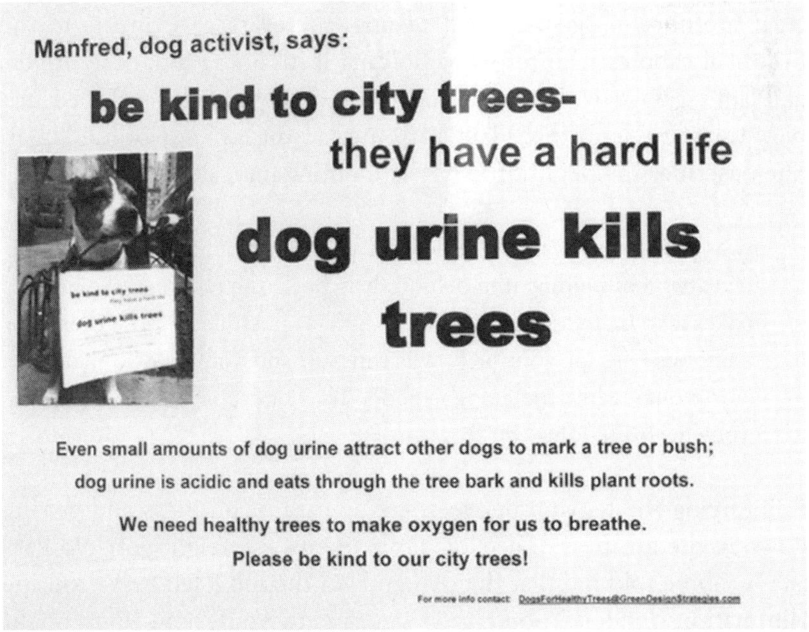

Figure 5.5. "Dogs Kill Trees" sign, New York, 2008.

high concentration of urine quickly kills small shrubs, flowers, and grasses and over time can strip bark from trees, leaving them susceptible to infections and pest infestations.[35] Biologically, then, piss and shit on the street pose a greater risk to the trees than to any other urban inhabitants. Dogs are innately wired to leave scent where scent has been left before, scattering it around instead of letting it all go at once, say, in the gutter. This results in a suffocating volume of nitrogen-rich urine trickling onto each tree and its surrounding soil. Even worse for the tree, the urine of small dogs has a higher concentration of nitrogen than other urine does and is thus especially toxic to trees. Even when humans must resort to peeing on the street, they are much less likely to pee on a tree than in a protected back alley or darkened doorway. So when it comes to the city's "public" health, the trees are at greater risk than are humans or pets, and their biggest threats are peeing dogs, even though official sanction condones the dogs and their owners while targeting human street excretors with summonses and fines.

Whereas human public health is only a spurious rationale for the legal prohibition against street peeing, there are real threats to the health of cabbies who resort to holding it in. Some longtime drivers at Plaut's garage with many years behind the wheel developed urinary-tract problems that Plaut attributed to the strategy of holding it. She describes the plight of a sixty-something man she calls Ricky:

> I was pretty sure Ricky's bladder and kidney problems were related to years spent holding it in behind the wheel. The nature of the job makes it so that you want to stop as little as possible because eventually business, or your luck, will run out, and you need to make your money while the going is good. Tons of cabbies have kidney problems from doing this.[36]

Reinforcing the tension between taking care of business and getting down to the business of driving a cab, one twenty-eight-year-old Pakistani cabbie told me that the demands of the job itself make routine elimination difficult. "Some guys get greedy, you know. They might have to go, but they just see another fare and take that fare instead of stopping for a break. And if that fare takes you out, you know, you might be driving out to Brooklyn, stuck on the bridge, and then have to come all the way back. Or out to the Bronx, and you don't know anywhere to go, so you just come back." He continued, "You know, holding your urine all the time, it's not good. Not good for health. Lots of cab drivers, they work for too long, they get diabetes." Although there is no evidence to support a relationship between driving a taxi and developing diabetes or the incontinence Ricky suffered, it is worth noting that even though this driver reported having five different places to go spread around Manhattan, he still found the lack of bathrooms to be a "big problem," big enough to generate a hypothesis among himself and his colleagues that too many years spent holding it in causes debilitating health problems.

There is also a gender dimension to the health concerns—women menstruate, which generates a need to stop that truly cannot wait. Women are also more likely to suffer urinary-tract infections than men are. Half of all women will have a urinary-tract infection in their lifetime, and women who have one infection are more likely to have another.[37] Symptoms include the persistent urge to urinate, painful urination, abdominal pain, and occasionally fever—conditions that make cab driving nearly impossible. Successful treatment requires prescription antibiotics, but driving a cab is not a job that comes with health insurance. Some evidence suggests that holding it in is indeed a risk factor for developing a new infection for those who have previously been infected, compounding the complications for women, who are anatomically more susceptible in the first place.[38] The monthly demands of menstruation and possible urinary-tract infections may stop some women from even considering becoming drivers, but they are more likely to become an insurmountable hurdle after time spent on the job.

Dogs Doing Boundary Work

Dog urine on the street is socially acceptable because dogs are not constrained by human systems of order and disorder. In fact, their act of freely peeing helps mark that boundary between order and disorder, between restrained civility and wild nature. The matter out of place created by humans who pee on the street is not the urine so much as it is the flagrantly free body and behavior of the urinator. The act is out of order, a social transgression whose costs are both symbolic and quite real, at least when the transgressors are arrested and/ or fined for their behavior. The difference between urine released on the street and urine deposited into toilets is negligible in the long run because New York's sewage system is designed to collect both public and private flows for processing at municipal water-treatment plants before release into New York's waterways.[39] Roughly half the time when it rains the single-stream sewage system is overwhelmed, and the overage is diverted directly into the waterways without treatment (additional capacity is decades in the making). Whether the sewage comes from the street or a fully plumbed toilet, it has an equal likelihood of being untreated before release, decreasing any physical distinction between street and private deposits of excrement.

Dogs and their owners work the boundary between the civilized and the uncivilized on New York's streets. The presence of peeing and pooping dogs on the street is a carefully scripted performance. Dogs are not simply incontinent. Manhattan's dogs spend the bulk of their days indoors, where most of them are able to control themselves for long hours while their owners are at work. In other words, although it is possible to train most dogs to go only in the gutter, for example, or never on the grass, they are generally let loose on the entire streetscape. The collective recognition of the public street as the right place for dogs to go casts them as guards at the boundary between human and animal, civilized and wild. As they freely pee, soaking the city in urine and sticking their noses in it, they help construct "wild" nature in an urban space where nature is a construction more than a circumstance. In urban space, our green spaces spring from the programs of landscape architects, and our trees are caged. Dogs privately restrain themselves most of the time in order to perform nature's call in public.

Rather than opening up possibilities, the freedom of dogs reinforces symbolic danger for cab drivers and others who are left with few socially acceptable places to go. If ever they answer the call of nature on the street, they risk falling on the wrong side of the human/animal binary. It is no small matter to embody too fully the animal "them" instead of the human "us." For mostly immigrant cab drivers, racism, xenophobia, and classism already threaten to cast them far from the ideal of the financially and socially affluent type of person riding in the backseat.[40] Struggling as it is to maintain dignity within the trying conditions of being treated robotically, another kind of nonhuman status, the drivers are more sensitive to the symbolic transgression of peeing in public than are the Saturday-night drunks, who can explain away their behavior with a shrug and a guilty grin.

The unashamed attitude of the dogs reminds us that what seems like the barbaric behavior of the city's honking, spitting, cursing human population is still somehow civilized, if not exactly civil, when cast next to the utter incivility of a squatting dog. As Elias reminds us, the civilizing process has occurred only "slowly and laboriously," starting with the control of the body, elimination and nudity in particular.[41] As early as 1530, Elias found writings that compelled "well-bred person(s)" to "avoid exposing without necessity the parts to which nature has attached modesty." Elias also found that later writings inched the civilizing process along by recommending modesty even when a person is alone because "angels are always present."[42] What was once the watchful responsibility of the angels is now thought to be embedded in one's sense of social belonging. The problem with peeing in public in the absence of any particular members of the public is not that the angels might see but that the person peeing will bear the weight of the shame of social transgression whether or not there is anyone else present. For Foucault, the shame of private peeing in the public sphere is produced by an authoritative gaze and experienced through bodies quietly guided by deep scripts for organizing perception.[43] A breeze wafting over exposed adult genitalia inspires not a child's squeal of delight at a body unhindered but a sense of vulnerability at being out of bounds, offensive: of being dirty. Genitals exposed in public are matter out of order, indeed.

The symbolic shame of transgressing the human/animal binary by answering the call of nature in the public gaze is not experienced by the transgressor alone. Revisiting Elias's civilizing process shows that it is not only individuals who become civilized over the life course but also whole social fabrics whose collective achievement of civilization is based on the behavior of its constituent members. When people pee freely in public, it is a problem for all New Yorkers, suggesting that we are collectively barbaric, unable to provide reliable provision for one of the fundamental steps toward civilization. From Elias's account, the precept that was first introduced in 1731, "If you pass a man who is relieving himself you should act as if you had not seen him," was restated more broadly by Erving Goffman in 1971: "A rule in our society: when bodies are naked, glances are clothed."[44] Thus, the onus of constituting a civilized society falls on all members. A relaxation of shock and shame is often afforded the very young and the mentally ill since these types of person are allowed to be outside the realm of the civilized without polluting the pot for the rest of us. But when social transgressions are made by those who are otherwise deemed fully, successfully socialized, the weight of the offense is not theirs alone to bear. When folks prefer not to pee in public, it is a shame on us, not a shame on them, if they are forced to do so.

The Pee or the People?

Leona Helmsley's decision to leave her substantial estate to dogs created a stir, perhaps in part because it occurred in a city already made a bit uncomfortable by its rabid inequality.[45] While pets are fed, bathed, trimmed, and treated like members of the family deserving quality time, humans obviously in need are often treated more like street furniture—objects or encounters to be navigated. Ethnographer of street-based booksellers in Greenwich Village Mitchell Duneier found that bathroom access was a key point of friction between his informants and the larger social setting on Sixth Avenue. One of his informants, who had been hassled by the police for public urination, was keenly aware that the problem was not the pee: "Even dogs, they go up and pee against the side of the building. If they have so much against humans peeing against the side of the building, they shouldn't let the dogs pee against the building."[46] The experience of being treated as lower than the dogs, socially out of order, is frustrating and alienating for those with nowhere to go. In a 2006 ethnography of a mixed-income, mixed-race neighborhood, Gabriella Modan echoes Duneier's findings. In this Baltimore neighborhood, public urination, she says, "is demeaning for the people who have to eliminate like animals (and even the animals probably have someone to clean up after them). This perpetuates the feelings of hopelessness and low self-esteem."[47] Exercising the practice enjoyed by dogs comes dangerously close to being a dog. For what makes us human if not our ability to control our animal bodies?

The legal distinction between urination across species makes clear that the problem is not the pee. Public urination is not a crime against public health but a violation of symbolic order that contributes to the social construction of both class and gender. In Alexander Kira's otherwise assiduously technical examination of the bathroom (see Penner, chapter 11, in this volume), Kira, quoting Reginald Reynolds, writes the following about bathrooms and bathroom behaviors:

> One of the bases of the class, or caste, system is "that we ostracize and despise those who do the most necessary and unpleasant tasks with the least opportunity to keep themselves clean . . . thus wealth and cleanliness are the marks of idleness (and superior virtue), dirt and poverty being the insignia of labor."[48]

For taxi drivers and other street-based workers, being unable to use standard restrooms is therefore not just a physical inconvenience but also an offense against symbolic order. Women have long carried a greater responsibility for the active maintenance of cultural standards of purity, particularly with respect to the body (see Kogan, chapter 8, in this volume). A man can be, at least in certain ways, dirty and still be a man, but a woman who is dirty is intrinsically out of place and a sign of rot in the wider social order. Immodest and complicit in the production of dirt as she pees in public or into her own version of a portable urinal, she is a more subversive character. What sort of woman is a woman who continues to drive a cab, knowing that there's a real chance she will pee awkwardly, maybe even dribbling a bit in semipublic behind the wheel?

As it is, in the almost completely male taxi-driver work force, the offense is against both cleanliness and masculinity. Peeing in a car, no matter what catches the urine, is "dirty" by Douglas's definition of dirt; urine most definitely does not belong in a car. It can get loose either from the penis or vessel and defile the body, clothing, or upholstery. Furthermore, surreptitiously peeing while seated is an emasculated posture compared to the erect posture of peeing at a urinal. These are the assaults—one on the symbolic achievement of cleanliness and the other on the performance of appropriate masculinity. The achievement of civility and gender, always interlinked with each other, are again at risk.

Negotiating with a shopkeeper or restaurant owner over restroom access becomes an argument not just about physical relief but also for inclusion in the class of folks who uphold symbolic cleanliness. Being denied access is an exclusion from this class—castigation into the category inhabited by the freely peeing dogs, whose public peeing is acceptable because dogs are amoral, inherently incapable of actively constructing the symbolic order. Being told there's no bathroom for you is not so much about the bathroom as it is about not being recognized as "one of us" who deserves access to dignified enactment. Without public facilities as fallback, such workers are at the mercy of strangers who, without apology, do not always take them in.

Rest Stop

Trucker Bomb

FINDING PARKING AND turnaround space for a loaded semitrailer truck is so difficult that truckers fashion urinals out of empty gallon jugs for use in the truck to avoid spending time and effort finding a restroom. Highway rubbish cleanup crews in the I-90 corridor of the U.S. western states and in Canada along the Canadian Expressway find thousands of gallons of bottled urine in the tall grasses annually, sometimes just at the wrong moment. When heated by the sun or swiped by the blade of a lawn mower, the "trucker bombs" explode, showering unsuspecting maintenance workers with warm, stale urine.

In 2003 the state of Washington responded by imposing the maximum littering fine—$1,025—on anyone caught improperly disposing of a trucker bomb. Protecting maintenance crews is just one concern. Trucker bombs are generally filled while vehicles are in motion. First responders at the scene of a fatal single-vehicle truck accident in Utah found the driver with his pants around his knees, an open jug of urine on the cab floor.

Okay, one last time: This is not a urinal.

Get caught tossing a bottle of urine and you'll pay $1,025.
Fines for littering range from $103 to $5,000. Remember, Washington
State Patrol has eyes out for violators. (Not to mention their noses).

Litter and it will hurt.

REPORT VIOLATORS
866-LITTER-1

Washington State antilittering sign. (Courtesy of the state of Washington)

6

Creating a Nonsexist Restroom

Clara Greed

THE ACHIEVEMENT OF a nonsexist restroom requires immense cultural, attitudinal, policy, legal, and architectural changes. As Lewis Mumford put it, "you can judge the quality of a civilisation by the way it disposes of its waste."[1] However important as cultural artifacts of our civilization, public toilet design is a despised, outcast branch of architecture, just as it is a stigmatized subject of general discourse. Designing toilets has been compared by architects to "doing latrine duty in the Army."[2] Within large architectural firms, it is shunted off to the work cubicles of those low in the pecking order, not to be much discussed with clients or those higher up in the firm. It would be as though medical doctors similarly avoided talk and physical representation of embarrassing aspects of the body—something which modern medicine had to overcome to be effective in its mission. As Sue Cavanagh and Von Ware draw the parallel,

> To a good doctor there is no physical or mental aspect of his [sic] patient which should embarrass him. He may be worried or shocked by what his diagnosis reveals, but if he's any good, he is not embarrassed. Correspondingly, therefore, there should be no type of building, and no human function related to it which should embarrass the architect.[3]

Without embarrassment, the first part of this chapter will investigate what the problem is, and the cultural and attitudinal reasons why it is not being solved. In the second part, I make recommendations as to how to create a nonsexist restroom, with respect to the different levels of the problem—at the macro, meso, and micro levels. These recommendations involve new directions for design and planning for the city as a whole, the local area, and the toilet block, respectively. Although I draw heavily on British toilet research (some of it my own), many of the issues are internationally applicable, as I have learned through my work with the World Toilet Organization (the other WTO).

What Is the Problem?

I came to investigate public toilets through my more general research on the social aspects of urban planning. When I asked ordinary people, especially women, what is wrong with our cities and towns, they frequently mentioned a basic lack of public toilets. People said that the public toilets that are available are sited in unsavory locations, are dirty, and often are inaccessible, especially for those with strollers or baby carriages. At the city level men have more places to relieve themselves. In Britain (and in many other European countries) it is common to find a public toilet block or street urinal for men but nothing for women. Within office buildings, factories, and other workplaces there are "normally" more toilets for male employees than for women, especially in erstwhile male-dominated occupations. The great de facto alternatives are bars, pubs, clubs, and licensed sports facilities, some of which bar women as a matter of policy or at least more subtly communicate signals of unwelcome. In the case of bars and pubs, children are not permitted as a matter of law, and hence their caregivers are excluded as well.

At the individual restroom level the disparity is the starkest. Even if the floor space provided for the ladies' and men's rooms is of equal size, men are likely to have twice the places to pee because they will have stalls plus a row of urinals along the wall—hence the queues, where women patiently stand in line. Women, for biological reasons and because of modes of dress, take twice as long to urinate as the average man.[4] Women have extra biological reasons for needing public toilets, such as pregnancy, menstruation, and a higher level of incontinence in old age. Women are more likely to be accompanied by babies, small children, or elderly relatives, all of whom need the toilet more often and need more space within the stall to accommodate the multiple humans and their equipment, as well as comprising a large proportion of the people with disabilities.

Why Has the Problem So Persisted?

The general embarrassment in society in talking about toilets and bodily functions undermines and shapes all aspects of toilet design, provision, and discussion. The very terms commonly in play—*restrooms, comfort stations, public conveniences* (to name a few of them)— are redolent with cultural embarrassment. Americans say they are "going to the bathroom" when they are heading to a room wherein, curiously, there is no bathtub. Similarly, Brits says they are about to "go to the loo"—from the French term for water, *l'eau*. Other euphemisms are also associated with water and washing, such as *lavatory*, or when people say they will "go wash up" or "wash my hands," activities that occur *after* the event that is the actual purpose of the trip. The terms putatively obscure the "dirtiness" of the activity.[5] In the United Kingdom, excretion is also covered over by a gloss of social class. The British speak of going to "the ladies'" or "the gents'," terms to confound the rudeness with the society of high and mighty. The term *away from home toilets*, as coined by the British Toilet Association, covers the relevant forms of public toilet, at least those of interest to me, and escapes some of the silliness.[6]

The social and ecological problems of equating excretion with dirtiness are not universally prevalent—or at least not consequential in the same way. In some societies, "brown gold" has been spread on the fields as an essential component in agricultural fertility. In Japan, annual ceremonies were held in some Shinto temples to celebrate such events, with the residue offered to the gods, the equivalent of the Anglican Church's Harvest Festival. In ancient Rome the emperor Vespasian charged people for the right to empty and clean the public toilets, because the waste was highly valued as fertilizer. The collection

Figure 6.1. Men's urinal built in an earlier, more luxurious, bathroom era.

of excreta for agricultural use appears across many anthropological accounts of farming technique. Urine is used not only as fertilizer but also, linked to its antiseptic qualities, as remedy for insect and snake bites as well as wounds from other sources.

The provision of toilets, in their modern and Western form, was once seen as an honorable professional career, indeed an intrinsic part of a valiant fight against epidemics. It had taken something of a transformation in public thinking to make the connection between disease and excreta, somehow overcoming Victorian sensibilities in avoiding such intimate issues. Once in force, however, provision of public toilets became conjoined with a nineteenth-century spirit of civic pride. It was an important component of "gas and water socialism" and "sewers and drains" modernization movements of the time. For an example of a fine standard for the public environment resulting from these movements, see figure 6.1.

But from the outset, public toilet provision for women was seen as an extra, as a luxury, or as problematic in other respects (a set of issues taken up by Kogan, chapter 7, in this volume). The main concern of the male city fathers was to provide toilets for men, whose role in public space was accepted and indeed regarded as important to the industrial economy.[7] And besides, men's capacity to pee on streets and walls stands as a threat if alternative facilities are not at hand. In recent times, to deal with men urinating while waiting for London night buses at Trafalgar Square—eroding the Portland-cement foundation of the National Gallery—authorities installed street urinals nearby (fig. 6.2).[8]

In contrast, women's right to urinate has not been as easily accepted, certainly not in the open and often not in any sense of public space. It is deeply embarrassing to men—perhaps even the fact that women do urinate at all, much less that they do so where others know it is happening. The lack of toilet provision for women, as many feminist writers have long argued, was no oversight but part of systematic restriction of women's access to the city of man.[9] That women, especially working women, nevertheless had to use the city created great suffering for them as well as for the more occasional female city users.

The belief in zoning to separate different land uses was a key theme governing the development of modern town planning in the nineteenth and twentieth centuries. It aimed to organize cities to make them sanitary. The fact of women's bodies, with menstruation as crucial aspect, risks polluting sites of the legitimate citizen. Sweeping women off the streets fit into this need for categorization: manufacture versus residential, male versus female. Women's presence threatened as well the maintenance of public/private boundaries, a fundamental element of necessary separations.[10] Women who move around outside the home, especially if it is out in the city versus, say, in a rural village, are "out of control," a phrase significantly also used of the prostitute and for both the urologically and morally "incontinent" (a word with many meanings). There is no female flâneuse equivalent to the flâneur, no "woman about town."[11] The "bladder's leash" tethers women to home through the lack of adequate public facilities.[12]

Why, given the obviousness of the problem, has solution been so hard to achieve? Members of the built-environment professions, such

Figure 6.2. UriGienic pop-up urinal (a) stored underground, (b) emerging from the ground a bit, (c) nearly completely emerged from the ground, (d) completely above ground, doors open, and (e) completely above ground, doors closed. (Courtesy of UriLift)

as architects, planners, surveyors, and engineers, are not value-free and certainly not gender-free.[13] Rather, the design and policy choices made by an urban decision-maker reflect that professional's life experience and assumptions of what the built environment should be like, and the position of women and men therein, along with the shared values and assumptions of the professional subculture to which "he" belongs. To change the toilets we need to change the professional culture of toilet providers. One way of doing this is to get more women and other so-called minorities into the construction, planning, and architectural professions in the hope that this will also result in a wider range of life experience and alternative viewpoints and priorities being introduced to policymaking.[14]

Women constitute over 52 percent of the UK population and the bulk of public toilet users, but they constitute less than 10 percent of toilet providers, that is, the professionals, designers, and managers responsible for determining toilet policy. In particular, there are very few women in some of the more practical ends of the relevant professional practices—the plumbing and engineering specialties. An obsession with structural and technical issues predominates over social, ergonomic, health, equality, accessibility, and livability issues, with women's needs peripheral. This feeds into a fortress rather than access approach to restroom provision. One sees this with architectural form as well as placement on the land. It gives rise to features such as vandal-resistant doors, absence of mirrors, and severe fixtures hardened against abuse. In some places in the United Kingdom, the loo lighting is ultrablue so as to make it impossible for addicts to see their veins. This may or may not affect rates of drug addiction (it may just increase the danger of using needles improperly), but it does cast a terrible depressing light over the environment of all visitors. Budget goes into the hardware rather than human attendants, who are more versatile in the services they could provide and softer in the way they could provide them.

Fear of windows (and their possible breakage) inhibits good ventilation. Making trade-offs that yield steps, even if only a few of them, especially disadvantages those with baby carriages or people who cannot move about easily. Putting public toilets down under the city, whether accessed by stairways, as was common in the United Kingdom at prior times (see fig. 6.3a), or via dark and dingy ramp (fig. 6.3b), as in the more recent era, was conceived without people's well-being foremost in mind. On the inside, fixtures and fittings have too often been based on the perceived needs of the healthy young male (with narrow hips), carrying nothing more than a rolled-up newspaper. He uses a urinal, designed and deployed for maximum efficiency—albeit with little concern for possible male anxieties over privacy, as Ruth Barcan explains in chapter 2 in this volume. A urinal, no matter how efficient, does women little good, forcing them to rely on the stall to relieve themselves. To get into a typical stall, one needs to do a three-point turn to get into position, something men have to do far less frequently.

Men do use the stall for sexual activity, and it is this role of the cubicle that seems to bring great concern among those responsible for managing the facilities. Because of such "cottaging" (as it is called in the United Kingdom) and other acts of concern, a senior toilet manager responsible for the fate of a whole city's toilets told me, "the only good public toilet is the closed public toilet." The women's side of the toilet is typically closed as well because of "problems in the gents." Traditional municipal on-street public toilets in the United Kingdom were closed at a rate of 40 percent between 1998 and 2008 (fig. 6.4).

When the fact that women end up with less public toilet access than men do is raised at toilet meetings (usually by me if I'm there), some pleasant young man is always likely to say, "Well, they could always use a pub." But besides the problem of children, pubs are not always open (especially in the morning), and women can be put off by their sometimes macho atmosphere. Many ethnic-minority and religious women do not enter any premises serving alcohol, nor would they use the toilet in a fast-food restaurant that is not halal. And as public toilet provision declines, more and more pubs and fast-food outlets—perhaps as defense from an increasingly needy populace—display notices to ward off noncustomers. In big cities, in part as a result of terrorist-related security concerns, you can no longer just walk into a hotel, office block, or other public building to use the toilet. Also increasing at office buildings are keypad access systems that require insider knowledge of which combination to tap.[15]

Figure 6.3. Handicap-inaccessible bathrooms (a) in the transit system and (b) unpleasantness in central city.

Figure 6.4. Shuttered restroom.

Legal Challenges

Many of the problems of today, at least in the United Kingdom, stem from legislation of the past. The main toilet act in Britain is still the 1936 Public Health Act, section 87, subsection 3, which gives local authorities the right to build and run on-street "public conveniences." It also authorizes them to charge such fees as they think fit, "other than for urinals," which must be provided for free. It is thus deeply discriminatory. This act is also flawed in not making it compulsory for local authorities to provide toilets; the powers are only permissive. With regard to gender spatial allocation, British Standard BS6465 for "Sanitary Installations" provides for men to have a third to a half more provision than women. As a result of my getting onto the revision committee, along with colleagues architect Michelle Barkley and enlightened engineer John Griggs, I have been able to change these ratios to provide higher levels of provision for women. But the standards are not retroactive. Putting them into effect often runs up against other standards, such as those for health and safety, fire regulations, workplace regulations, and so on. It means that a quick and inexpensive fix is problematic. The whole situation needs a complete reworking with a comprehensive approach to all the various important design criteria.

Recent trends in equality legislation and public consciousness give hope. The 2006 Equality Act requires the British government to implement the new Gender Equality Duty (the GED) as a result of European Union directives. The GED stipulates that all providers of public goods and services must prepare a Gender Equality Scheme that gender-proofs all aspects of public policymaking, resource allocation, and personnel management. As with past problems of implementation, local authorities and government departments have been slow to realize that public toilet provision is a key factor to be evaluated from a gender perspective, with appropriate seriousness and priority. So far, sadly, local government toilet providers have seen the GED as a means of enabling them to charge men as well as women, rather than increasing the level of provision for women. Or rather than fulfill its terms, or those of the previously legislated Disability Discrimination Act (1995), requiring access upgrades for the disabled, some local authorities have simply closed all their public toilets—a time-honored tradition of depriving everyone.

What Is Needed

Macro Citywide Level

For an urbanist strategy, we need to look first at where to locate the toilets, because many a good toilet has been put in the wrong location and been underused, never found, or left vulnerable to destruction. We have inherited a hodgepodge of toilets which were built at different moments for reasons lost in the mists of time. Patterns of location (and gender ratios) are linked to outdated travel patterns, related to male work and leisure activities including drinking and sports. Especially since large numbers of women both work and care for others, they are more likely to undertake multiple errands in a single trip (trip-chained journeys, as planners call them), multitasking as they go, rather than monopurpose commuter trips (even men now are less likely to follow the classic commute journey). A typical daily trip-chain for a woman might be from home to the childcare provider, to school, then on to work, then back from work, to the school, to the childcare provider, perhaps to the dentist, to the shops for food purchases, and then home. Women working in local or adjoining neighborhoods, or out on industrial estates, are more likely to have tangential than city-center-focused journeys, creating an alternative complex web of trips. A citywide toilet strategy should be part of the wider process of mainstreaming gender into urban planning policy, with an entire rethink of zoning, transportation, and locational policy in the light of the "discovery" of women's different travel and activity patterns.[16]

Following from a survey of all toilets to which the public has access both on street and off street and disaggregated by gender, there should be a citywide hierarchy of toilet provision, with the highest level of provision in central areas but adequate provision in district centers and in areas where more localized needs exist. Rail and bus station restrooms need to be planned as part of the overall transportation strategy, rather than as afterthoughts to that planning. In certain places, such as shopping malls, women can constitute as much as 80 percent of those on the premises. Such hot spots need to be taken into account in the determination of ratios, a kind of site-specific planning that requires some subtlety to bring off correctly.

A toilet strategy needs to take into account time as well as urban space. Some sites shift in the gender (and age) ratio over the course of a day, a week, or the calendar year. A facility that serves attendees at a wrestling match can be put to service for a fashion show or children's puppet theater. The football stadium might be taken over by a pop festival or an evangelistic rally. Drinking happens at night, and the results hit at particular hours; it used to be at pub closing time of 11 p.m. in Britain. In many parts of the world, local authorities have been introducing street urinals for male drinkers, with no alternative solution for women. Some of these are designed to rise up from the ground at night, but then disappear in the early morning (see fig. 6.2). On challenging various officials on the sexist nature of providing only urinals, which flies in the face of the Gender Equality Duty, I have been told that it is not discrimination, as these are mechanisms that safeguard all (women as well as men) from the unpleasantness of street fouling (twenty-four-hour toilets for everyone would be a better solution).

Meso District Level

At the neighborhood and district level of provision especially, it is helpful and feasible to survey existing facilities to detect pressures and conditions as well as to conduct surveys of the local population to determine needs. This can happen as part of larger investigations of desired facility and infrastructure changes at the community level—surveys of the sort I have taken part in conducting in the United Kingdom. In this way, my colleagues and I have assessed users' needs and travel and activity patterns and attempted to disaggregate the information on the basis of age and gender. Children, quite obviously, need toilets located in parks and play areas, as well as in shopping precincts—where their needs can be very immediate and where the facility thus must be close at hand. Of course, facilities need to be in schools (the dire tale of school toilets is another matter that deserves close attention in its own right). Elderly women have told us they want toilets nearby to post offices and libraries and that every cemetery should have accessible toilets for widows and widowers visiting their spouses' graves. Some people want restrooms big enough to bring in their dogs and bicycles, too.

Long ago women planners recommended the introduction of "multipurpose" community buildings in suburban residential areas, especially in the United States, where the suburbs are so extensive in scale.[17] Working mothers would be assured that a range of childcare, elder care, and after-school care could be provided locally, along with public toilets. The fact that the restroom is often built "on its own," as an outcast among buildings, may be a spatial reflection of anxiety and, indirectly, sexist and elitist attitudes toward those who most depend on public toilet access. Public toilets need bringing in from the cold; they should be seen as features of public art in their own right, not clutter to be swept away.[18] Let's have fabulous, glamorous toilet architecture. This is why I have always seen toilet provision as part of urban design and of sustainable, livable, accessible cities and mixed-use vibrant streets, à la Jane Jacobs.[19]

The research done by my colleagues and me also revealed that people want toilets to be located alongside major roads and in car parks at their destination. A key issue that came up in our interviews is the need for at least some basic minimal provision in less populated, peripheral, and rural locations.[20] Indeed, having such facilities, and having them at the proper toilet intensity, can help even a small sleepy village come alive once a week when a fair or market might take place. Even if it does not lead to permanent installations of entertainment or amusement, an outcome that may not be desirable in any event, it would help create liveliness on a regular basis, a boon to life in nondense areas.

The "interval factor" also should be taken into account. Aggregate daily demand is obviously not a sufficient basis for standards at locations where a lot of people need access during a short space of time, such as school breaks, theater intermissions, sports halftimes, and so on. Distant beauty spots—national parks, seaside locations—which are usually tranquil may suddenly be swamped when ten buses arrive all at once, full of predominantly female senior citizens all wanting to use the toilet. Street drinking (and urinating) is particularly a problem in Northern European countries which have a beer-based drinking culture (the lager lout scenario) as against the wine culture of Southern Europe or the higher permitted drinking age in the United States. Beer involves inebriation through volume, and when beer drinking is done by many in a small space of time, it creates a surge of need at specific hours and places. As with sewers and flood defenses, it is important to build for maximum possible volume, not the minimal or average. Solutions include a more flexible approach of converting men's rooms to women's facilities (or vice versa) or creating the facilities that can be used by either sex—a vexing moral, legal, and architectural challenge taken up at some length by other authors in this volume (Mary Anne Case, Harvey Molotch, Olga Gershenson).

Integrating restrooms with other buildings—such as the local tourist center, information kiosk, or café—might reduce vandalism and save costs. Restrooms can also be conjoined with post offices, libraries, and other public buildings. It might be particularly advantageous to design them into police stations and firehouses; such buildings are open day and night anyway and provide reassuring environments for those concerned about rapists or other potential miscreants. Of course, that would draw those otherwise charged with guarding public safety into the realm of intimate human service, perhaps threatening their own sense of (mainly) manly dignities. Maybe that is the reason this has not been done as widely as makes sense. It could also be arranged for only the women's room to be so situated, with the men's room placed—especially if costs or some other practicality made it difficult to locate both genders within the same building—at a location farther afield.

It is quite remarkable that the so-called New Urbanism, probably the most influential force in contemporary planning and development of new towns and rehabilitation of old ones, has virtually nothing to say about facilities for human elimination.[21] The whole movement is based on an embrace of neighborhood, walkability, and "return" to civic consciousness. It advocates, in its principles, "increased density," "quality of life," "interactivity," and greening in all sorts of ways. But provision for humans to go out and relieve themselves is not on the list. New Urbanism should discover old necessities.

Micro Detailed Level

Given women's disproportionate needs, it is clear that a solution for women would be helped by simply more provision in general. There are many debates as to how many toilets are needed per person in the population. The British Toilet Association has recommended that

> a local authority should provide no fewer than 1 cubicle per 500 women and female children and one cubicle and one urinal per 1,100 men, and no fewer than one unisex cubicle for use for people with disabilities per 10,000 population and no fewer than one unisex nappy changing facility per 10,000 people dwelling in the area.[22]

The relevant "population" should also include commuters, tourists, and visitors as well as residents. But it is crucial to tailor facilities to the actual people who use them. Again, the best solution is to include survey work, basing provision on what real women and men indicate that they want.[23]

No matter the subtleties involved, in virtually all cases, authorities must allocate substantially more space to women. This would happen in the United Kingdom, given the sad state of current affairs in some places, if only a one-to-one ratio of parity were achieved. Take a look at the layout of men's and women's rooms of the Glyndebourne Opera House in southern England, built in the year 1994 to a fine standard but with completely inadequate facilities for women (fig. 6.5). A two-to-one ratio in favor of women is what is necessary, but even in such a deluxe facility (and one obviously frequented by at least as many women as men) that ratio is nowhere near being provided. Laws requiring this ratio (or something close to it), already on the books both in Britain and in many U.S. states and municipalities, do not deal with already-existing structures. Given that few new facilities are being built, only retrofitting already-built structures would appreciably improve conditions for women. In Japan, to cite one example where such standards are taken seriously and facilities have in many cases been retrofitted, queues are no more at the ladies' room. One solution, one that addresses temporal changes in demand between women and men, is to design a mechanism for adjustments to be easily and cheaply made, much in the way that hotels routinely and quickly divide ballrooms into smaller meeting halls. Especially in an all-toilet environment, room partitions could shift along with gender changes over the course of the day or week. This is not an insurmountable design problem; it just has received little attention.

Inside the facility, the layout and dimensions of the toilet cubicle are a crucial issue for women. Let's start with the door. If it is inward opening, it reduces the space for the user to turn around and sit on the seat—not an easy maneuver, as previously mentioned, in a compact space (fig. 6.6). I have found examples where the edge of the door is touching the front edge of the toilet bowl. There should always be a reasonable gap between the two of at least 250 mm, ideally

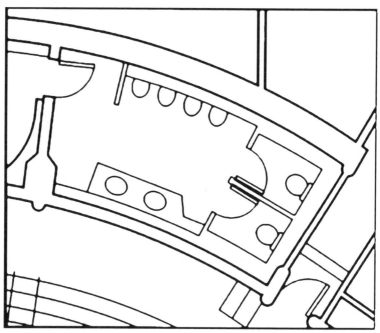

Figure 6.5.
Glyndebourne Opera bathrooms (a) for men (above) and (b) for women (below).

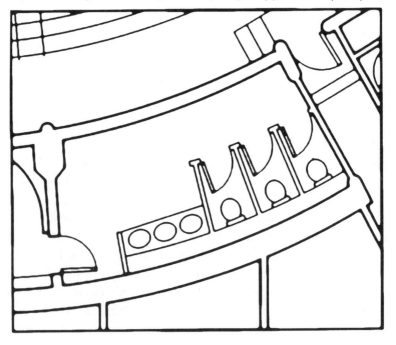

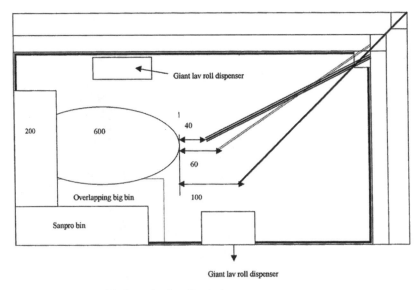

Figure 6.6. Diagram of the hazards of small cubicles.

more. Again, keep in mind that women may be struggling with packages, large purses, childcare equipment, and the wriggling children themselves (fig. 6.7). The high gap between floor and door (as well as at stall sides), presumably to ward off unseemly behavior between men in the gents', has been thoughtlessly applied to the women's toilets too. Women need privacy, not surveillance.

There should be adequate space to sit on the seat without rubbing against the walls. In Britain there is often a giant toilet roll on the sidewall above the bowl that further intrudes. Then, on the other side, there is likely to be a large plastic bin for the disposal of sanitary towels and tampons (fig. 6.8). This frequently touches or overhangs the seat, making it difficult to sit. The solution is simply a cordon sanitaire of 450 mm around the bowl where nothing can be placed. Disposal facilities can be integrated on the back wall. The present arrangement presumes that menstruation is such a minor issue that it can be dealt with by a temporary plastic bin rather than as an integral component of toilet plumbing and waste disposal—and without the need for additional space provision within the stall. The bins arrived as add-ons, the result of misplaced

Figure 6.7. Babykeeper hanging pouch for keeping babies close in the bathroom. (Photo courtesy of Mommysentials)

environmental legislation that classified sanitary waste as clinical waste. No one bothered to work out the space implications within an already-crowded toilet compartment. Given the fact of menstruation, access to water for cleansing (as in a bidet) would have special benefit for women. It would also help users from those parts of the world where people use water, not toilet paper as the method of cleansing.

Some women who do not sit on the seat use paper seat covers when they are provided (particularly if, as is common in Japan, they dispense automatically). Lacking cover provision, women may use toilet paper to cover the seat before sitting, with further risks to drains and ecological waste. A toilet for squatting would bring benefits to women, especially to those who are not actually sitting anyway. A nonsexist restroom (as well as nonorientalist) might thus include the choice of at least one squat toilet with accompanying provision of water supply in the cubicle (fig. 6.9). For those who use

Figure 6.8. Crowded bathroom stall.

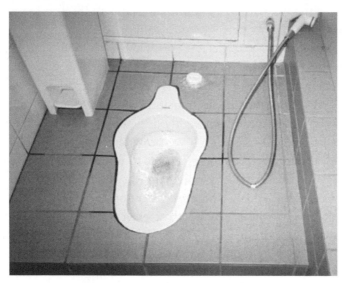

Figure 6.9. Paper-free squat toilet, Islamic world.

toilet paper, the tissue should be clean and easily get-at-able, rather than squashed into some impenetrable plastic contraption that won't unroll—or a metal container that miserly dislodges one small square at a time. Many women use toilet paper to blot themselves after urination, and so it will touch one of their most intimate areas. No wonder many women carry their own paper and tissues. No toilet paper or provision of harsh scratchy paper (as still found in the United Kingdom as well as other parts of the world) is clearly sexist, as women need soft paper more than do men. For similar reasons, women pay a harsh penalty if hygienic soap, hot water, and decent drying capacity are absent. Having all such facilities within the cubicle itself, in the manner of the airplane toilet (see Case, chapter 10, in this volume), greatly aids in dealing with menstruation as well as care of children, including those sick with diarrhea or nausea or bleeding from minor accidents.

Albeit with shades of "Big Brother" and risks from malfunction and breakage (see Braverman, chapter 4, in this volume), automatic systems for opening and closing the cubicle, flushing water, and operating the tap do help—and again they especially help women, given their routines and responsibilities. The APC (automatic public convenience) is the ultimate in an automatic setup and is also the ultimate unisex facility. Too often, however, authorities replace a traditional row of toilets with a single APC. APCs have a long washing cycle, and queues move slowly. Even large versions make little allowance for strollers and shared (parent and child) occupation or for dogs and luggage. They provide no transition area between the toilet facility and the street, something highly desirable no matter what type of toilet is involved. A transition foyer provides a psychological (and/or physical) shelter when queuing, a place to wait for friends or a helper or to leave a baby carriage. Women are wary of toilet doors that open straight onto the street, where they cannot see what they are stepping out into. A simple spy hole (or even a large panel of one-way glass—see "Rest Stop: Flirting with the Boundary," in this volume) would also help, as found in some Asian toilets. Surveillance cameras mounted outside the toilet but with visibility from within would put the otherwise-worrying technology to work on behalf of anxious women.

Debates rage as to whether there should be breast-feeding facilities within toilets or at least provided as a set-aside part of the facility. In an ideal nonsexist world, nursing would take place in the open in any comfortable location, but in the meantime some provision away from men's prying eyes seems appropriate. But breast-feeding should be acoustically and visually insulated from other toilet functions; most of us do not want to eat where we (or others) defecate. Mothers may want no less for their babies.

My colleagues and I have argued from our research that having toilet attendants cuts down on vandalism and that their presence is valued by users. With attended toilets, the design can be more open and accessible. There should be due consideration to implications for cleaning (by the attendant or whoever is responsible). Cleaning up other people's waste, the role assigned to slaves and untouchables, should be assigned—as much as possible—to the fixtures and architectural layout. Hence toilets mounted on walls allow hands-free mopping underneath, and at least elements of the APC should be applied, however much feasible, to do the dirty work. Again, there is a gender angle: in many parts of the world—India comes to mind—the cleaning of both men's and women's toilets is assigned to women.

My suggestions all involve changes in cities, buildings, maintenance, staffing, and fixtures that adapt to women's needs. The alternative, of course, is to change the women. An example is the idea that women should learn to use a device for vertical urination, as described by Mary Anne Case in chapter 10 and in this book's introduction. These things are awkward to use and present problems of sanitation and portability. They should be resisted. Similarly suspect are calls for women to stop "wasting" time in the restroom, allegedly primping and socializing. Putting aside the empirical validity of such allegations, which I have reason to doubt, it represents another attempt to "blame women" rather than face the costs, both to male privilege and in public finance, that humane reform would require.

Conclusion and An Essential Asset or a Drain on Resources

My suggestions do involve spending money, and it is a question as always of where the money will come from. Should the provision of more public toilets be self-financed, perhaps by advertising sales— what we might regard as the neoliberal solution? Increasingly, where public toilets can still be found, one has to pay to get in—the APCs being especially consistent and rigid in this regard (imploring the machine for a free entry will do you no good). The alternative is for authorities to regard restrooms as a public good and make them free. Given women's heavy needs, this would be a boon to them. Put the other way round, restroom turnstiles exact a sexist "tax."

Municipal public toilet providers constantly complain that there is no money for toilets, while finding funds for much-criticized public art installations, sports facilities, convention halls, and various other extravagant initiatives. It is all a matter of priorities; and therefore it is a matter of the cultural perspective and life experience of the decision-makers themselves. Public toilets are one of the few concrete benefits that women, the elderly, and those with disabilities really value. Why aren't we even angrier? From my frankly feminist perspective, patience, trust, and obedience to the powers that be will never result in urinary equality. At the micro level, I like to indulge in a little toilet evangelism when I am standing in the queue for the ladies' with a captive audience, in spite of the looks I get. To paraphrase Marx, many women are suffering from false toilet consciousness. Why isn't it a major political issue? And conversely, why doesn't the government take toilets seriously? What happened to the civic pride manifested in previous generations? (See figs. 6.1 and fig. 6.10.) Why aren't we "splendid in our public ways"? Much of the answer comes down to sexism and to the fact that many politicians, decision-makers, and managers have very little idea of ordinary people's lives; we are invisible to them.

It is also plausible, although it should not be necessary, to argue for public toilets as a public benefit on other grounds besides the direct human satisfaction they deliver. If there is adequate toilet provision, people will stay longer in the city; they will spend more money

Figure 6.10. Grand bathroom built in a previous era of bathroom pride.

on shopping, entertainment, and as tourists. The citizens will be healthier and cost the employer and health providers less money. The sustainability case works well too. My colleagues and I have argued from our research that if the government wants to get people out of their cars and back onto public transport, walking, and cycling, then public toilets are the missing link. They need to be present at transit nodes and where pedestrians cross paths. People cannot be expected to use public transport unless they can be secure in the comfort and health of their bodies—and that means having access to public toilets at strategic sites along the way. Particularly for women, it comes down to toilets, and all human life—urban and otherwise—is there.

Rest Stop

A Woman's Restroom Reflection

Be Prepared for the Women's Room, by Anonymous
(Who Is, Reportedly, a Woman)

When you have to visit a public bathroom, you usually find a line of women, so you smile politely and take your place. Once it's your turn, you check for feet under the stall doors. Every stall is occupied.

Finally, a door opens and you dash in, nearly knocking down the woman leaving the stall.

You get in to find the door won't latch. It doesn't matter, the wait has been so long you are about to wet your pants! The dispenser for the modern "seat covers" (invented by someone's mom, no doubt) is handy, but empty. You would hang your purse on the door hook, if there was one, but there isn't—so you carefully, but quickly, drape it around your neck (Mom would turn over in her grave if you put it on the floor), yank down your pants, and assume The Stance. In this position your aging, toneless thigh muscles begin to shake. You'd love to sit down, but you certainly hadn't taken time to wipe the seat or lay toilet paper on it, so you hold The Stance.

If you   sprinkle
when you tinkle
Please be neat
and wipe the seat -
& even if
you're in a rush,
please don't forget
to flush!

"If You Sprinkle When You Tinkle" sign in women's restroom of an architectural office, New York, 2007.

To take your mind off your trembling thighs, you reach for what you discover to be the empty toilet-paper dispenser. In your mind, you can hear your mother's voice saying, "Honey, if you had tried to clean the seat, you would have known there was no toilet paper!" Your thighs shake more. You remember the tiny tissue that you blew your nose on yesterday—the one that's still in your purse. (Oh yeah, the purse around your neck, that now you have to hold up trying not to strangle yourself at the same time.) That would have to do. You crumple it in the puffiest way possible. It's still smaller than your thumbnail.

Someone pushes your door open because the latch doesn't work. The door hits your purse, which is hanging around your neck in front of your chest, and you and your purse topple backward against the tank of the toilet. "Occupied!" you scream, as you reach for the door, dropping your precious, tiny, crumpled tissue in a puddle on the floor, lose your footing altogether, and slide down directly onto the toilet seat. It is wet, of course. You bolt up, knowing all too well that it's too late. Your bare bottom has made contact with every imaginable germ and life form on the uncovered seat because you never laid down toilet paper—not that there was any, even if you had taken time to try. You know that your mother would be utterly appalled if she knew, because you're certain her bare bottom never touched a public toilet seat because, frankly, dear, "You just don't know what kind of diseases you could get."

By this time, the automatic sensor on the back of the toilet is so confused that it flushes, propelling a stream of water like a fire hose against the inside of the bowl that sprays a fine mist of water that covers your butt and runs down your legs and into your shoes. The flush somehow sucks everything down with such force that you grab onto the empty toilet-paper dispenser for fear of being dragged in, too.

At this point, you give up. You're soaked by the spewing water and the wet toilet seat. You're exhausted. You try to wipe with a gum wrapper you found in your pocket and then slink out inconspicuously to the sinks.

You can't figure out how to operate the faucets with the automatic sensors, so you wipe your hands with spit and a dry paper towel and walk past the line of women still waiting. You are no longer able to smile politely to them. A kind soul at the very end of the line points out a piece of toilet paper trailing from your shoe. (Where was that when you needed it?) You yank the paper from your shoe, plunk it in the woman's hand, and tell her warmly, "Here, you just might need this."

As you exit, you spot your hubby, who has long since entered, used, and exited the men's restroom. Annoyed, he asks, "What took you so long? And why is your purse hanging around your neck?"

This is dedicated to women everywhere who deal with public restrooms (rest? you've got to be kidding!). It finally explains to the men what really does take us so long. It also answers their other commonly asked questions about why women go to the restroom in pairs. It's so the other gal can hold the door, hang onto your purse, and hand you Kleenex under the door!

7

Sex Separation

The Cure-All for Victorian Social Anxiety

Terry S. Kogan

THOUGH ONE CAN find examples of sex-segregated water closets in public spaces in the United States well before the end of the nineteenth century, the first law mandating such separation was enacted by the Massachusetts legislature in 1887.[1] That statute, entitled "An Act to Secure Proper Sanitary Provisions in Factories and Workshops," required that "suitable and proper wash-rooms and water-closets shall be provided for females where employed, and the water-closets used by females shall be separate and apart from those used by males."[2] By 1920, forty-three states had adopted similar legislation.[3]

What was the perceived need in late-nineteenth-century America for the strong arm of the law to intervene to enforce a social practice already developing informally? This chapter argues that policymakers were motivated to enact toilet separation laws aimed at factories as a result of deep social anxieties over women leaving their homes—their appropriate "separate sphere"—to enter the work force. Although women had been leaving their homes since the early century to work in factories, especially textile mills in the Northeast, by the end of the century the anxiety over women in the workplace was fueled by and became conflated with other social anxieties: concerns over the fast growth of technology and the related dangers of injury; concerns over public health that resulted from the devastating cholera epidemic during the Civil War; and Victorian concerns of privacy and modesty, especially related to men and women working side by side in American factories. It was the confluence of these anxieties that led to an urgent sense that the law needed to step in.

To understand how toilet separation laws came about, it is important to trace the rise of these anxieties over the course of the nineteenth century and to examine the distinctive architectural solution adopted by policymakers to allay these concerns: sex separation of public spaces.

Women and Social Anxiety in the Late Nineteenth Century

At the outset of the century, both men and women worked at home, a space devoted not only to rearing the family but also to retail business and professional practice.[4] Beginning early in the century, however, the economic restructuring of American society as a result of the industrial revolution led to a "division of spheres."[5] The home ceased to be the central economic unit as men left for new public workplaces where manufacturing was centralized.[6] As a result, an ideological division between public space and private space emerged during the course of the century; the workplace became the domain of men, the home that of women.[7] Accompanying this division of space arose a new vision of women and their role in society:

> During the nineteenth century, the genteel elite—as well as an emergent middle class—developed an ardent faith in the civilizing power of moral women. Females were widely assumed to be endowed with greater moral sensibility and religious inclinations than men. Such pedestaled notions of women helped nourish a powerful "cult of domesticity" which assigned to women the role of self-denying guardians of the hearth and soul. As the more complex economy of the nineteenth century matured, economic production was increasingly separated from the home, and the absence of men who left to work long hours in the city transformed the middle-class home into a "separate sphere" governed by mothers.[8]

The ideology that enshrouded the home as the separate sphere of women has been dubbed the "cult of true womanhood," a vision that infused antebellum America.[9]

The problem with the sentimental vision of the virtuous woman entrenched in her domestic sphere is that it was a myth, bearing little resemblance either to women's daily experiences or to the evolving social realities of the nineteenth century. From its outset, the century witnessed women leaving the privacy of their homes and entering the public world of the workplace.[10] Beyond the workplace, women also moved into the civic life of the community, becoming active in social reform and suffrage movements.[11] Moreover, the presence of women in sites of social engagement such as theaters and public parks, though previously the domain of men, became accepted by the second half of the nineteenth century.[12] In addition, new, semipublic spaces began developing with women patrons in mind, including the first American department store, opened by A. T. Stewart in New York in 1846.[13] Nonetheless, any move by women outside the domestic sphere was viewed by many people with serious concern, for the growing number of women in public spaces evidenced a "living contradiction of the cult of true womanhood."[14]

Independent of concerns over women leaving their homesteads, social anxiety over public health was also heightened in the second half of the nineteenth century. Antebellum America was a filthy place.[15] Because public-works systems capable of delivering water to private homes were not constructed in most cities until the late 1870s, few homes had running water. With the exception of very wealthy people, homes did not have indoor bathrooms as we know them today. Even among the better off, "despite the growing bourgeois devotion to sanitation in person and in the kitchen, the outdoor privy was still the norm in polite society."[16]

The American public health movement was brought about largely by cholera epidemics, rampant disease, and death during the Civil War. Prior to that time, disease was considered a "scourge of the sinful."[17] Only after postwar development of the germ theory of disease did Americans begin to understand that sickness was brought about not by human moral failure but by a lack of sanitary conditions. There-after, sanitation became recognized as a science, and Americans began to take hygiene seriously. Established in the 1870s, a public health movement led by reformers known as "sanitarians" dedicated them-selves to discovering the underlying principles of "sanitary science."[18] Armed with that knowledge, they began attacking the haphazard self-contained plumbing arrangements that had developed in Ameri-can homes. By 1890 extensive public waterworks connected private homes to municipal water systems, and crusaders began lobbying for passage of plumbing codes to standardize household plumbing.

Finally, anxiety pervaded late-nineteenth-century America as a re-sult of the extraordinary growth of technology and industrialization, perhaps most notably in the areas of transportation and the growing presence of factories. As a result, every facet of individual life seemed invaded by the noise and perceived dangers of new technologies, in-cluding streetcars and railroads. This rapid growth of technology threatened the ideological divide between public space and the private home, resulting in a deepening sense of anxiety in Victorian society.

In the face of this anxiety, personal privacy became a major concern to urbanites. "The right of individual privacy, under new pressures in the brashly inquisitive metropolis and subject to the development of new technologies of intrusion and publicity, was elevated to sacred status, which everyone was bound to respect."[19] Associated with the newly developed interest in privacy, late-Victorian society became ob-sessed with concerns of modesty, concerns surrounding the human body and bodily functions.[20] These concerns were deeply intertwined with issues of social morality.[21]

The Rise of the Realist Movement

To appreciate how policymakers responded to these social anxieties and to understand what led to the adoption of laws mandating that factory toilets be separated by sex, one must consider the rise of the midcentury intellectual movement known as "realism" and the im-pact of that movement on social attitudes toward women.

Historians have characterized the first half of the nineteenth century as an age of sentimental idealism. "Idealists shared a basic conviction that fundamental truths rested in the unseen realm of ideas and spirit or in the distant past rather than in the accessible world of tangible facts and contemporary experiences."[22] During this period American culture displayed a "sovereign disregard of reality."[23]

The second half of the nineteenth century witnessed the development of the realist movement, a movement fueled by the rise of science and its commitment to "verifiable knowledge and tangible concerns."[24] Considered by its advocates to be a rejection of early-nineteenth-century idealism, realism infused every aspect of late-century intellectual and artistic life. Legal scholars have long recognized the influence of realism on classical legal thought, a movement in legal theory in the second half of the nineteenth century that sought to align law with the rise of the sciences.[25] Among other features, classical legal thought tended to view the world in physicalized, spatial, and boundaried ways.[26] As demonstrated later in this chapter, this approach to the world led scientists and policymakers to direct the focus of regulation on physical bodies and architectural spaces.

Despite realism's seeming commitment to the scientific method, there was nothing neutral about the realist approach to gender. Realists "viewed 'things out there' through a lens of confining social conventions and moral inhibitions. Considerations of the marketplace, class consciousness, racial and gender prejudices, and deeply embedded standards of morality and decorum often narrowed the borders of the realistic impulse."[27] Early in the nineteenth century, differences between men and women were a matter of folklore, theology, and philosophy. After midcentury, the understanding of gender shifted from a focus on idealized social roles to a focus on physical bodies.[28] Though science had long explored the differences between men and women, late-nineteenth-century sexual science aspired to be more empirical than previous inquiries into the nature of gender had been, calling on the new social sciences of anthropology, psychology, and sociology.[29] Scientists in these disciplines reached the common conclusion that "women were inherently different from men in their anatomy, physiology, temperament, and intellect."[30] What previously had separated men from women were matters of etiquette and proper social roles. What now separated men from women were innate biological differences, a conclusion that bolstered the validity of the separate-spheres ideology.[31]

Rather than abandon the early-century vision of the vulnerable virtuous woman protected in her homestead, realist scientists and scholars found "facts" to vindicate that vision. Armed with these hard facts, realist policymakers directed their regulatory energies at women's bodies and the public spaces inhabited by those bodies.

The Realist Regulation of Architectural Space to Protect Women in the Public Realm

Given the realist focus on bodies and spatial boundaries and given the unwavering commitment to the separate-spheres ideology, mid-nineteenth-century planners had a ready solution for dealing with anxieties over the emergence of women from the homestead: re-create a home-like, separate sphere for women within the public realm itself. Thus, beginning at midcentury one finds numerous examples of architects and planners cordoning off spaces in public buildings and other accommodations for exclusive use by women. Such separate spaces were created in railroad cars, commercial photography studios, department stores, hotels, restaurants, banks, post offices, public parks, and libraries.[32] These spaces were often designed to reflect decorative details of the home. An examination of several of these spaces sets the stage for understanding the direction that lawmakers eventually took in enacting laws mandating that factory water closets be sex separated.

Ladies' Reading Rooms in Public Libraries

A rarity in America before 1850, the few public libraries that existed were bastions of male status that often excluded women. As public libraries began to develop, the question of women's presence became a serious issue. Some library leaders advocated admitting women into public libraries to assure that private libraries would continue to be exclusively male enclaves. Others, however, were concerned that women would be disruptive to the concentration of serious readers.[33]

Nonetheless, embracing the vision of the cult of true womanhood, many library leaders believed that women would enhance a library's cultural mission to uplift the populace. But women's moral superiority also led such library leaders to perceive them as vulnerable to the advances of vulgar males.[34] The solution to allowing women into public libraries was architectural: create a separate ladies' reading room and then stock it with fashion and home advice magazines. In 1859, the Boston Public Library opened its first ladies' reading room, located on the floor below the general reading room. By the last quarter of the nineteenth century, a separate women's space became an accepted part of American library design. The furnishings in such rooms were generally less institutional than those in the rest of the library, often reflecting the furnishings in a private home. One common feature of such rooms was a hearth, which combined with the furniture, carpets, and window treatments to reflect the domestic spaces associated with women's separate sphere. To protect the modesty of "true women," these rooms often provided discreet access to the women's toilet, invisible to other parts of the library. Abigail Van Slyck explains:

> Ladies' reading rooms established in American public libraries in the late nineteenth century did not welcome women as full participants in the public sphere. Rather they played an active role in reproducing a particular set of gender assumptions. Their design and location suggest that they constituted a partitioning of the public sphere through the provision of specially arranged settings that encouraged female readers to assume culturally prescribed postures of genteel femininity.[35]

The "Ladies' Car" on Railroads

The growth of transportation technology was considered especially dangerous to women.[36] Beginning in the 1840s, American railroads began designating a railroad car for their exclusive use, known as the "ladies' car." The spatial significance of the ladies' car did not stop with the simple creation of a separate space. The car was generally placed at the end of the train, which "spatially reflected men's obligation to protect women's physical safety," since those nearest the front of the train suffered the greatest injury in the event of a crash.[37] Moreover, the distance of the ladies' car from the engine assured that it had the cleanest air. In contrast, men were relegated to smoking cars, the atmosphere of which was more like a tavern or men's club, a place of smoking, chewing tobacco, and drinking. Respectable women rarely ventured into the smoking car. Other spaces related to railroad travel were also re-created to offer separate, special accommodations for women, including ticket windows and station waiting rooms.

The Ladies' Parlor

In addition to libraries and railroads, a women-only parlor space was created in a range of other establishments, including photography studios, hotels, and big-city department stores.[38] "This separation," as Katherine C. Grier explains, "was not absolute; women could and did participate in mixed-sex social life in the saloons of steamboats, on railroad coaches, and in large parlors of hotels. . . . The ladies parlors were islands of domesticity in the realm of otherwise unregulated public life."[39] Grier concludes, "Ladies' parlors in particular provided a domestic haven which buffered their interactions with the broader public in commercial spaces, as well as providing an interim solution for a young society which was still looking for a 'place' (in both the psychological and the spatial sense) for women in the public sphere."[40]

During the third quarter of the nineteenth century, other public spaces set aside for exclusive use by women included drawing rooms, dining rooms, and ladies' entrances in hotels; ice cream parlors; a ladies' parlor in New York City Hall; and a ladies' window in San Francisco's post office.[41] These spaces accomplished little in the way of actually protecting women and their "weaker" bodies from danger. Rather, while acknowledging the ever-increasing presence of women in public, these spaces reinforced the cultural message that, as the weaker sex, women needed special home-like havens when they ventured into the threatening public realm. If forcing women back into the home was not a realistic possibility, planners settled on the alternative of re-creating aspects of the domestic sphere in public architectural design.

Protective Legislation Aimed at Working Women

For several decades, policymakers were content to allow architectural and design professionals to use sex separation as an informal tool to address concerns over women in the public realm. Harnessing the law to enforce such separation first occurred late in the century with respect to the factories and other workplaces. Why?

Though the early-century separate-spheres ideology viewed women as vulnerable when they left their domestic havens, its idealistic nature did not lead to the enactment of protective legislation. Moved by the scientific pretensions of realism after midcentury, however, legislators began to take seriously the threat that allowing vulnerable women into the public realm would endanger both women's weaker bodies and the welfare of future generations. More than any other public venue, the factory became a flashpoint around which the heightened social anxieties that infused late-nineteenth-century society converged. First and foremost, it was the destination to which women were perceived as going when they left their homes. In addition, factories often contained dangerous machinery, chemicals, and jobs, dangers to which "weaker" women and their potential offspring were deemed to be particularly vulnerable. Moreover, despite the added attention given to public works and sanitation in the broader society, little attention had been paid to factory sanitation by the end of the century. Conjoined with the physical dangers and poor sanitation in factories, concerns arose over the moral perils that faced women who worked in close proximity with men.

Several decades before lawmakers adopted laws mandating sex-separate factory toilets, they began adopting general labor legislation aimed at protecting working women. The move to enact protective labor legislation in the United States came early in the nineteenth century in response to the radical transformation of daily life wrought by the industrial revolution. Though attempts were made to limit the working hours of all workers as early as the mid-1820s, such legislation proved ineffective because it allowed the employer and employee to agree contractually that such limitations would not apply.[42]

Moves to adopt protective labor laws aimed more narrowly at working women and children proved more successful. In 1852, Ohio adopted the first law limiting the hours that women would be allowed to work in factories.[43]

Thereafter, states began enacting laws aimed directly at protecting the health and safety of women workers, including laws prohibiting women from engaging in certain professions or work assignments deemed dangerous,[44] laws restricting or prohibiting night work by women,[45] laws mandating that women be given relief time for meals,[46] laws mandating a rest period during the working day,[47] laws prohibiting the employment of women immediately before or after childbirth,[48] and laws aimed more generally at protecting a woman's reproductive capacity.[49] Laws requiring that seats be provided for women workers in a wide range of industries were adopted in virtually every state.[50]

Early protective legislation was often justified based on the inherent weakness of women's bodies and on the potential danger to their reproductive capacity.[51] For example, a study by Grace F. Ward for the Women's Educational Industrial Union in 1909–10 stated,

> Legislation in the past has recognized that conditions of labor for children and women are very closely allied by nature. Both classes are admittedly in need of greater protection by the public than is usually afforded to the working man, and this for cogent reasons. The physical strength of the working man is less likely to fail through overwork; and even where it does fail, the effect on future generations is less serious than a similar deterioration in the mothers.[52]

Irrespective of the justification, such legislation generally achieved protection for working women from very real dangerous conditions in factories, protection that would have been welcomed by all adult workers.

The move to direct the strong arm of the law at factory sanitation and toilet facilities came only later in the century. In fact, such regulation could not have been enacted until the 1870s, when public-works technology first enabled plumbing to be brought indoors and effluence to be transported though municipal water systems. Prior to that time, factory toilets consisted most often of outdoor privies or outhouses.

Despite the intense focus of sanitarian reformers on public health beginning in the 1870s, they were slow to direct their attention to factory sanitation. Prior to the advent of sanitary science and the germ theory of disease, concern over factory sanitation focused not on dirt but on dust; discussions in the scientific literature concerned ventilation.[53] The cause of unhealthy conditions in the factory was deemed to be "atmospheric vitiation," the inhaling of impure air. Mention of toilets and sanitation was generally in the context of inadequate ventilation.[54]

Apart from the poor sanitary conditions of factory toilet facilities, they were generally not separated by sex. Women and men could alternate use of the same commode and thus be brought into close proximity through such common use. An 1887 report by New York factory inspectors makes this clear:

> The workshops occupied by those contracting manufacturers of clothing, or "sweaters," as they are commonly called, are foul in the extreme. Noxious gases emanate from all corners. The buildings are ill smelling from cellar to garret. The water-closets are used alike by males and females, and usually stand in the room where the work is done.[55]

Little change in either sanitary conditions or sex separation had occurred by the early twentieth century. In a section entitled "Toilet Accommodations," a 1914 New York investigation into factory sanitation notes,

> No part of an industrial establishment is so neglected as the toilet accommodations. In many cases they are located outside of the factory, and sometimes quite a distance from it, causing the loss of much time and also endangering the health of the employees.
>
> In the investigation made for the New York State Factory Commission, the toilets were located in yards in 186 of the establishments inspected. . . . Many of the toilets were not separated for the sexes and were of an obsolete and crude type.[56]

The first laws requiring that factory water closets be sex separated were adopted by states as amendments to and extensions of earlier protective labor legislation aimed at women workers; these laws were not intended as neutral regulations for the mutual benefit of men and women alike. Though the first such law, adopted by Massachusetts in 1887, was not directly linked to prior protective legislation,[57] when New York became the second state to adopt a separation statute aimed at factory toilets two months later, it did so by explicitly amending earlier protective legislation aimed at women.[58] Other states adopting such laws followed New York's lead by also amending existing protective legislation aimed at women.[59] The very title given such legislation often disclosed a clear legislative intent to protect women (and at times children), not the entire working population. For example, the 1897 Tennessee law first requiring that factory toilets be sex separated was entitled "An Act to Require Employers of Females to Provide Separate Water-Closets for Them." Similarly, the 1919 North Dakota law adopting a sex-separation requirement was entitled "An Act to Protect the Lives and Health and Morals of Women and Minor Workers."

Laws Mandating Sex-Separated Factory Toilet Facilities as a Cure-All for Victorian Social Anxieties

In contrast to earlier protective labor legislation aimed directly at alleviating dangerous workplace conditions, it is not at all obvious what led regulators to conclude that separating factory toilet facilities by sex would protect working women. Insight into this issue can be gained from reviewing both the numerous turn-of-the-century reports that examined the conditions of working women and the toilet legislation itself.[60] A review of this literature suggests that sex separation was viewed by late-nineteenth-century regulators as a kind of cure-all for the full range of social concerns surrounding factory women.

As was true of earlier legislation, legislation aimed at workplace sanitation was justified in part as necessary to protect women's weaker bodies. For example, a 1903 Department of Labor study that considered the effects of "insanitary conditions" in factories stated, "Women suffer even more than men from the stress of such circumstances, and more readily degenerate. A woman's body is unable to withstand strains, fatigues, and privations as well as a man's."[61]

Women's greater physical vulnerability led to the recommendation that separate, special facilities for women should be provided in the workplace. A 1913 report on sanitary conditions in the dress industry states,

> There is, however, one important matter of sanitation in which the shops suffer in common with the shops of many other industries, namely, the absence of lunch and retiring rooms. In the shops where there are a large number of girls working, it is probable that there are a number likely to have sudden attacks of dizziness, fainting or other symptoms of illness, for whose use provision should be made in the form of rest or emergency rooms.[62]

Accordingly, that report's final recommendations suggest not only that "a separate water-closet apartment shall be provided for each sex" but also that "in all shops where more than twenty-five women are employed, a provision shall be made for rest and emergency rooms for their use."[63]

Other justifications offered in the literature for sex-separated facilities ventured far from women's weak bodies to broader social concerns pervading Victorian society. By the late nineteenth century, the sanitarians and their commitment to "scientific plumbing" had gained considerable prestige and broadened considerably the scope and inclusiveness of the sanitary reform they sought.[64] The recommendation that factory restrooms be sex segregated became conflated by sanitarians with other requirements of sanitary science related to piping, water supply, or sewage. This is made clear in the writings of J. J. Cosgrove, a highly regarded sanitary engineer who published a number of technical books and histories used in colleges and technical schools to teach sanitation and plumbing architecture.[65] In his 1910 treatise, *Plumbing Plans and Specifications,* Cosgrove devotes a section to "Planning the Plumbing for Factory Buildings," in which he states,

> So far as the comfort and convenience of the employees are con-
> cerned all factory buildings are very much alike and not any of
> them require a great amount of plumbing. Of the small amount
> that is required a supply of drinking water will be found in the front
> rank. . . . Outside of the drinking fountain all that will be required
> are separate toilet rooms for the men and women and another toi-
> let room for the office help. Ordinarily it will be found that one
> water closet and one urinal for each 20 male employees, or part of
> that number, and one more closet for each 20 female employees, or
> part of that number, is the smallest possible allowance.[66]

For Cosgrove and other sanitarians, the sex-separation requirement
was indistinguishable from other requirements of sanitary science,
such as sufficiency of number of toilets, cleanliness, lighting, and ventilation.[67] (Also worth noting is the blatant lack of parity between the
number of facilities designated for men and the number designated
for women, an issue addressed, with vigor, by Clara Greed in chapter
6 of this volume.)

Deeply influenced by sanitarians, legislators similarly collapsed
and conflated the statutory requirement that factory toilets be sex
segregated with other, more technical requirements of sanitation.
For example, an 1889 California statute appended a sex-separation
requirement to a mandate that factories and similar establishments
"shall be kept in a cleanly state and free from the effluvia arising from
any drain, privy, or other nuisance, and shall be provided within reasonable access with this efficient number of water-closets or privies
for the use of the persons employed therein."[68]

In addition to justifications related to protecting women's weaker
bodies and to furthering factory sanitation, requiring sex-separated
water closets was also justified as necessary to protect the virtue and
modesty of women in the workplace. This justification was often embodied in statutes and reports using language concerning the "privacy" of toilet facilities. For example, the same section of the Connecticut law that mandated separate toilet facilities also required that
"every manufacturing, mechanical and mercantile establishment . . .
provide adequate toilet accommodations, so arranged as to secure

reasonable privacy, for both sexes employed or engaged in any such establishment."[69] Some statutes required that toilet accommodations for the different sexes have "separate approaches,"[70] while others required that such accommodations be "properly screened."[71] Connecticut's law further required that "when any such accommodations intended for use by any female adjoin such accommodations intended for use by any male, the partition constructed between such accommodations shall be solidly constructed from the floor to the ceiling."[72]

Given the Victorian obsession with privacy in the face of an increasingly intrusive industrial society, factories posed a special problem. Obviously, the "more intimate functions" had to be performed in such locations.[73] Statements such as the following make clear that concerns of privacy and modesty were of much greater concern with regard to women than to men:[74]

> In a very large proportion of the mills there is not reasonable privacy of approach to the water-closets. In some cases the water-closets for females immediately adjoin those for males. In some mills the construction of the water-closets is disgraceful; closets are built within the workrooms, and the thin board partitions do not extend to the ceilings, and in some instances the doors do not reach to the floor. Where this is the case the feet and lower parts of the skirts of females occupying the water-closets can be seen from the workrooms.[75]

It is a violation of Victorian modesty for any part of a woman's anatomy to be subjected to public scrutiny while she performs intimate bodily functions. The literature paints a picture of male workers defiling a woman's virtue by illicitly sneaking peeks at her lower extremities while she is using the water closet. Moreover, shielding a woman during such use was not enough. Victorian modesty was threatened if a woman could even be seen entering the facility.[76]

Justifications for sex separation of factory water closets based on women's weak bodies, factory sanitation, and modesty were all related to late-century anxieties that evolved from the realist movement's scientific pretensions concerning women's bodies and from disruptions to daily life wrought by a burgeoning industrial society.

But underlying these justifications was a deeper anxiety that pervaded nineteenth-century society from its outset, concern over the movement of women from the home to the public realm. As suggested earlier, realist science never shucked off its ideological commitment to the view that virtuous women should remain in their separate sphere. Accordingly, the literature concerning factory sanitation and the conditions of working women suggests a more basic justification for the sex-separation requirement: separating workplace water closets was necessary to protect and vindicate social morality. Mention of morality, though often intertwined with concerns of women's health, safety, and/or modesty, harked back to early-nineteenth-century ideological concerns of pure womanhood and separate spheres.[77]

Strong evidence as to the continuing vitality of the separate-spheres ideology is embodied in another of J. J. Cosgrove's books *Factory Sanitation,* published in 1913 by the Standard Sanitary Manufacturing Company of Pittsburgh, one of the country's major manufacturers of plumbing fixtures.[78] *Factory Sanitation* served two functions. The second half of the book consisted of an extensive catalog of workplace bathroom fixtures manufactured by the company. The first half of the book contained an extended essay also entitled "Factory Sanitation," by J. J. Cosgrove. His essay was intended to serve as a technical manual to advise companies about the most up-to-date ways to design factory toilet facilities. Accordingly, much of the essay is written in highly technical language. Nonetheless, Cosgrove also was interested in advising factory owners on how well-planned facilities could enhance employees' happiness.[79] The essay is illustrated with extensive pictures of both dilapidated and well-designed factory toilet facilities. Beneath a picture of a filthy wooden structure in the corner of a workroom (fig. 7.1) the following extended caption appears:

Toilet Facilities as Bad Morally as from a Sanitary Standpoint

Moral decency requires that where males and females are employed, separate accommodations shall be provided which, in every sense of the word, will be private. Ignoring the obvious filth of this double accommodation for "men" and "females," close

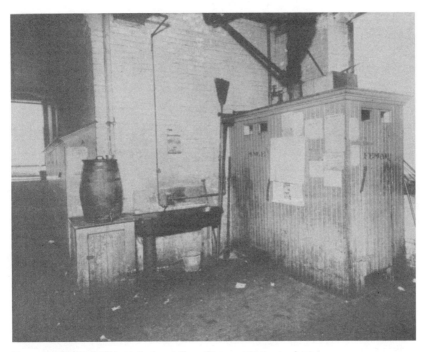

Figure 7.1. Toilet facilities as bad morally as from a sanitary standpoint.
(Courtesy J. J. Cosgrove)

proximity of the fixtures separated only by a thin board partition, far from sound proof, and the common approach, such accommodations would be morally objectionable even if they were sanitary, clean, well lighted and well ventilated.

Apply the golden rule in business. You would recoil with horror at the thought of your daughter being forced to avail herself of such accommodations. Treat other men's daughters, then, as you would like them treat yours.[80]

Though set forth in a technical scientific essay on factory plumbing and sanitation, Cosgrove's concern for a sex-separated bathroom is not founded in the requirements of sanitary science, for he notes that even sanitary facilities would be "morally objectionable" if not separated by sex. Rather, the appeal to one's "daughter" is meant to invoke the vision

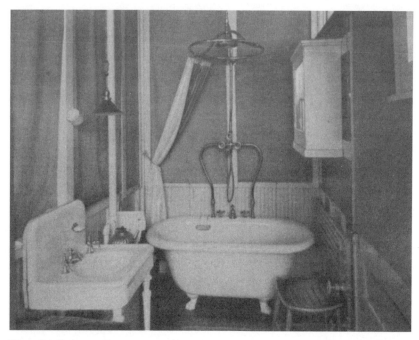

Figure 7.2. Bathroom for women in a Dayton Factory, Dayton, OH, c. 1933.
(Courtesy J. J. Cosgrove)

of the virtuous woman as pure and virginal. Despite the scientific pretensions of the realist and sanitarian movements, the moral ideology of
the early nineteenth century continues to shape technological decisions
and legislation at the turn of the twentieth century.

Cosgrove suggests elsewhere in his technical treatise that the sex-
separated toilet space for women should aspire to be home-like. Beneath a picture of a well-designed women's toilet room (fig. 7.2) the
following caption appears:

Bath Room for Women in Dayton Factory

Suggestive of all the comfort, cleanliness and convenience of a
bath room in the home. Can self-respecting capable operatives be
blamed for preferring such accommodations to those shown in the
front part of the book?[81]

Other experts on factory plumbing also associated the sex-separation requirement with early-nineteenth-century moral ideology. In an 1882 work on factory sanitation in Britain, B. H. Thwaite introduces a highly technical section entitled "Closet Arrangements" with the following:

> The importance of a proper closet accommodation, and its effects on the health and morality of the workpeople, especially in mills and workshops where operatives of both sexes are employed, will be acknowledged. Much immorality, vice and disease have been fostered by abominable closet arrangements. . . . The closets should be arranged in convenient positions. The closets for males should be distinctly separate from those for females.[82]

By the late nineteenth century, women were present throughout the public realm and played a major role in supporting the expanding American economy. That presence in factories and other workplaces, however, raised particular concerns about dangers to their health and, by extension, to future generations. At the same time, the early-century separate-spheres ideology continued to exert a strong cultural force throughout society. Given the confluence of concerns surrounding working women, beginning in the 1880s policymakers enacted laws mandating sex-separated toilet facilities in the workplace. These laws can best be understood as an attempt by legislatures to re-create the separate-spheres ideology within the public realm. If women could not be forced back into the home, substitute protective havens would instead be created in the workplace by requiring the separation of water closets, dressing rooms, resting rooms, and emergency rooms. In so doing, policymakers used a realist approach to legislation by manipulating architectural space to enforce social values.

Accordingly, despite common intuitions, the historical and social justifications for the first laws that required the sex separation of public restrooms were not based on a gender-neutral policy related to anatomical differences between men and women. Rather, these laws were adopted as a way to vindicate early-nineteenth-century moral ideology concerning the appropriate role and place for women in society.

Were the impact of this practice benign, the facts uncovered in this chapter might offer little more than satisfaction of a historical curiosity. The impact of toilet laws, however, is not benign. As discussed in more detail by other authors in this book (see especially the chapters by Clara Greed, Olga Gershenson, and David Serlin), laws mandating sex separation often create serious difficulties in daily life for transsexuals and transgendered people, persons with disabilities, parents with opposite-sex children, intersex persons, and women in general. Moreover, sex-separated public restrooms foster subtle social understandings that women are inherently vulnerable and in need of protection when in public, while men are inherently predatory. In addition, the ubiquitous presence of two restrooms in public places conveys the symbolic message that there are two, and only two, sexes, a message highly problematic to the public's acceptance of transsexual, transgender, and intersexual people.

Understanding the social and historical origins of separate public restrooms as evolving from a now discredited nineteenth-century ideology of pure womanhood and separate spheres can help free us to envision alternative arrangements for this architectural space, ones far more friendly to all persons who grace the public realm.

Rest Stop

MIT's Infinite Corridor, Now Shorter for Women

Sign indicating that a bathroom in MIT's infinite corridor will no longer be for men but will become a women's room after June 2005, Cambridge, MA, 2005. (Courtesy of Sam Maurer)

POSSIBLY NAMED BY a woman searching for relief, the "infinite corridor"—as it is generally called by people at MIT—is a spine-like, 825-foot, indoor pedestrian highway uniting the components of MIT's core academic campus. Until 2005, three of the four restrooms along this long central thoroughfare were for men. During a renovation of the corridor's east end, one of the restrooms underwent an architectural male-to-female transgendering. Urinals were removed and additional sinks and stalls added.

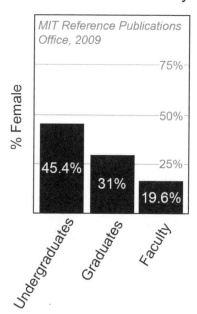

MIT Students & Faculty

MIT Reference Publications Office, 2009

Graph of the percentage of female students, faculty, and staff at MIT in 2008.

Women's acquisition of more equitable bathroom real estate followed reports on the status of women faculty that revealed unsettling gender gaps in overall representation, salary, and lab-space allocation. The architectural move is a statement in stone indicating the institute's increased sensitivity to gender equity.

8

Pissing without Pity

Disability, Gender, and the Public Toilet

David Serlin

IN APRIL 1977, a coordinated group of disability rights activists staged protest actions at the Department of Housing, Education, and Welfare in Washington, D.C., and in eight of its regional offices across the country. These demonstrators, many of whom used wheelchairs or mechanical ventilators, were fighting for the full-scale implementation of the Rehabilitation Act of 1973. Its significant Section 504, established to extend civil rights legislation of the 1960s, would prohibit programs that receive federal funding from discriminating against people with disabilities. Historically, people with disabilities had been either segregated within or isolated from the social world for so long that their public presence typically solicited pity rather than registering recognition. In San Francisco, disabled activists deliberately defied that history by occupying the Health, Education, and Welfare offices for twenty-five days, fed and cared for by their friends and attendants as well as by local unions, Bay Area countercultural groups, and Oakland's Black Panthers, all of whom recognized in the struggle of the disabled activists much of what other civil rights organizations, both radical and mainstream, had tried to achieve in the 1960s and early 1970s.[1]

Section 504 was an heir to the antidiscrimination laws that emerged from Lyndon Johnson's Great Society programs of the mid-1960s that were used to redress social, structural, and architectural manifestations of a segregated society. As articulated by activist Judy Heumann, "When you erect buildings that are not accessible to the handicapped, you enforce segregation."[2] For many disability rights activists in the 1960s and 1970s, the built environment could serve as an immediate conduit to social tolerance and inclusive citizenship. Such would be the benefits—physical and social, symbolic and material—of things like curb cuts, wheelchair ramps, and teletypewriter devices, along with transportation options such as kneeling buses. These accommodations were also intended to challenge the legacies of "separate but equal" treatment, which for the disabled was manifest as separate vehicles and spaces rather than vehicles and spaces that were integrated.[3] They are technologies of empowerment to inspire a more fully democratic society.

The lack of available toilet facilities during the occupation of federal offices in San Francisco did not prevent disabled activists from sustaining their protest but in fact may have shifted its terms, since their inability to use the toilet was both symbolic of and material evidence for their exclusion from the public sphere. As historian Joseph Shapiro has written, "Some of the most severely disabled protesters were literally putting their lives on the line, since they risked their health to be without catheters, back-up ventilators, and the attendants who would move them every few hours to prevent bedsores, or who, with their hands, would cleanse impacted bowels every few days."[4] Given the human need to eliminate, the decision to tolerate unpleasant and potentially lethal personal circumstances made the protesters' very presence a courageous act of civil disobedience.

By lifting the veil on their most intimate bodily functions, the demonstrators demanded recognition that the piss and shit of the disabled were produced not by androgynous bodies or amorphously asexual bodies but by bodies shaped by the same kinds of material and experiential needs as the able-bodied. Public provision for disabled people thus raises broader issues of what it means to recognize "special needs" in the context of claims, sometimes competing and complex, for dignity and equity. This understanding runs counter to the idea that providing, in minimal material terms, a place for the disabled to go to the toilet solves the problem.

How, then, do we begin to understand the disabled toilet as a technology that ostensibly neutralizes social difference if it exists within a public sphere that privileges able-bodied status—and a gender-normative able-bodied status at that? The premise behind the disabled toilet, after all, is that it transforms the public restroom into a level playing field; and once this field is so made, no one gets—or deserves—any further special privileges or consideration. Once there is compliance with the Americans with Disability Act (ADA), or some other regime of physical accommodation, the architectural and political deed is done. Yet this presumption of equality obscures the complex logic of the situation, at least in part because equality before the law is a paradigm that has been organized historically around male able-bodied privilege. As anthropologist Carol A. Breckenridge and philosopher Candace Vogler have written, "The 'person' at the center of the traditional liberal theory is not simply an individual locus of subjectivity (however psychologically fragmented, incoherent, or troubled). *He* is an able-bodied locus of subjectivity . . . who can imagine himself largely self-sufficient because almost everything conspires to help him take his enabling body for granted (even when he is scrambling for the means of subsistence)."[5]

The discourse of equality, then, is not always a sophisticated mechanism designed to generate respectful difference between parties with shared access to social or political power. Given what we know about the pressure toward assimilation experienced by many ethnic and sexual minorities in the United States, the discourse of equality also can be a blunt instrument used to flatten difference. As disability historian Henri-Jacques Stiker has argued, the discourse of equality for the disabled can be as oppressive as it is liberating, marking "the appearance of a culture that attempts to complete the act of identification, of making identical. This act will cause the disabled to disappear and with them all that is lacking, in order to assimilate them, drown them, dissolve them in the greater and single social whole."[6]

Much anxiety still remains at the core of contemporary encounters with difference, especially as the able-bodied must continue to grapple with tolerating the Other while valiantly struggling to defend their own status as the default position of public culture. And few sites, it seems, are as fraught with encountering the terms of difference and navigating the conditions of equality as that of the disabled toilet, as evidenced by the arcane rules of social etiquette and awkward confrontations that take place at the borders of tolerance and patronizing gesture. A dam of unintended consequences can break through the discomfited silence, as the "normal" person seeks ways not to notice, to avoid looking like they are trying not to notice, all the while busily defending their own capacity for privileged accomplishment.[7]

The disabled toilet in the United States emerged genealogically in parallel to other dilemmas of liberalism that continue to haunt discussions of difference. In the nineteenth and early twentieth centuries, the emergence of gender-segregated public and commercial spaces such as schools, gymnasiums, and bathhouses was closely linked to the influx of immigrants from Eastern Europe, Scandinavia, and the Middle East to large and medium-sized American cities. The example of the gender-segregated bathhouse did not by itself establish any legal or architectural precedents for disabled toilets, but it did make material the presumption among civic leaders that there was a demonstrable link between creating accessible public and commercial spaces and facilitating the hygienic and economic uplift of those at the margins.[8] Following the establishment of a public bath by the New York Association for the Improvement of the Condition of the Poor in 1852, free municipal baths, swimming pools, and showers for the residents of cities such as Boston, Chicago, Detroit, New York, Philadelphia, and San Francisco were considered politically and morally necessary by social reformers.

While public toilets became increasingly common throughout much of the nineteenth and twentieth centuries, people with disabilities were not among those populations around which the design and functionality of public toilets were organized. By contrast, able-bodied individuals were able to meet and exceed conformity to moral and bacteriological expectations of good health in the public sphere. Even today, people with disabilities, who carry the stigma of dependence and lack of control over their bodies, have often been perniciously associated with failures, deliberate or otherwise, of personal hygiene.

As the various social and political movements of the 1960s bore their fruit, authorities responded with national design standards for updating existing facilities or building new ones. The federal

Architectural Barriers Act of 1968 implemented the now-familiar curb cuts and ramps for wheelchair users. Along with the act came a host of proposals to make resources such as public toilets accessible, though the first innovations in toilet design, such as grab bars, lowered sinks, and so forth, were initially developed in the 1960s for private homes and institutions specifically established for the disabled rather than for public spaces. Only later did their installation in public toilets become reality, with the rules finalized in 1980. The new stalls, known as A117.1, were built at an increased depth of sixty-six inches, ten inches larger than the previous standards, and were to be built on a very slight incline from the floor so that wheelchair users could navigate and position themselves more easily within the stall, and ambulant but visually impaired disabled people could "feel" a slight step up that physically distinguished the stall area from the rest of the public toilet. The International Code Council's A117.1 standard also required metal grab bars behind and to the left side of the commode that could be used by the toilet patron or his or her assistant in order to help set the body in the appropriate eliminating position.[9] There were analogous modifications to existing sink designs to incorporate levers with large, wide paddles that can be moved easily, rather than faucets that demand greater muscle coordination and strength. These related standards of accessibility became law, in the form of the ADA, which brought with it a host of other legal mandates for equal access. It was not uncommon even in the 1970s and 1980s for disabled people traveling on public transportation such as trains or airplanes to have to wear adult diapers or use catheters and collection bags for the length of their journey due to inappropriate or nonexistent facilities.[10]

To a large degree, the exclusion of people with disabilities from the composition of the public, which went largely unchallenged until the disability rights movement of the early 1970s reached critical mass, grew from the stigma historically attached to numerous forms of visible disabilities in the public sphere.[11] It is true that veterans and survivors of at least some types of catastrophe had their injuries exhibited as examples of patriotic sacrifice. Their bodies, in the late eighteenth century and beyond, were usefully deployed rhetorically to promote national ideologies in times of crisis and war,[12] and the federal government acknowledged the specific needs of its paraplegic and amputee veterans from the Civil War onward. By contrast, many cities in the United States expanded nineteenth-century municipal vagrancy codes to prohibit people with "disgusting" bodies from appearing on city streets. As literary historian Susan Schweik has argued in her book *The Ugly Laws,* many of these laws were not lifted until well after World War II.[13] Those individuals who were not institutionalized or cared for at home or through local institutions or federally mandated resources such as the Veterans Administration were left to fend for themselves. Able-bodied personhood remained the default position, and the goal of rehabilitation was to integrate the disabled person back into society, but often with only primitive ideas and techniques for making it happen. So even wealthy disabled people who otherwise could design their own large and elegant bathrooms were forced to use chamber pots and other makeshift receptacles before the evolution of toilet designs dedicated to nonstandard or nonnormative body types, something that did not begin until the 1960s.

For much of the twentieth century, manufacturers of commodes, urinals, sinks, and faucets as well as the walls, doors, and floors that composed standardized public bathroom fixtures were oblivious to the body with special needs.[14] Innovations in toilet design lagged for a number of reasons—the social exclusion and disaffection of the disabled from all corners of public life, for one. But a second factor, perhaps ironic, was the ascent of mid-twentieth-century research sciences such as ergonomics, industrial design, and anthropometry that adopted a rather rigid and uncompromising sense of the "normal" physical body as the basis for design. This may have run parallel with the zeal for a generally conforming landscape with normative assumptions of physical ability as part of the iconic vision of an American nation. Any sort of distinctive body type perceived to be deviant was considered repugnant. The deployment of scientific authority and legislative barriers in the social control of difference is a testament to the power of putatively objective discourses that increasingly helped shape understandings of those who constituted the public, and those who did not, for much of the twentieth century.[15] Data from the U.S. military, based on young recruits, was a standard source in creating artifacts and appliances of many sorts, including New York City subway turnstiles.[16]

Disability itself comes in wide variations, and it takes close attentiveness to those variations to anticipate the range of potential needs. For people who use canes and crutches, for example, grab bars or roomier stalls may be a pleasing feature but ultimately an irrelevant one. On the other hand, grab bars on or near urinals may be a useful accommodation for an ambulant disabled man who is visually impaired or who uses a walker.[17] And blind people have very different sets of needs. For them, it is important that things be maintained in standard locations, something that can conflict with innovations designed to help those with other types of disabilities, as well as with the general public's desires for stylistic novelty and functional design improvement.

These complications raise the question, as can often come up in such discussions, of to what degree those with disabilities should strive to accommodate the world as it is and to what degree the world should change because there are disabled people in it. Put in a more positive way, what are the benefits to others of taking into strong consideration those with nonconforming traits? Many types of people, disabled or otherwise, have benefited from the ADA and accommodations that exceed it. Clara Greed, in particular, has emphasized how many of the design principles employed in disabled toilets—doors without handles, roomier stalls, lowered sinks, and interior spaces that allow one to move around other patrons—help those who do not identify as disabled, including the elderly, the temporarily disabled, large-sized people, and parents pushing strollers. Ramps and more commodious spaces help those traveling with suitcases or making a temporary pit stop while delivering goods.

On a more profound analytic front, the disabled toilet and its attendant controversies show us something more about the nature of "dependency"; they flag the possibility that autonomy may not, in itself, be without limits as a desirable social goal—not just for the disabled but for people in general. Autonomy often rests on a culture of deference without intimacy, granting others the "right of way" without having the patience for empathy. The people who encounter each other in public toilets, men especially (see Barcan, chapter 2, in this volume), act as independent agents tied together through little more than an ethos of benign neglect. At the same time, the homo-social space of the public toilet is, by its very nature, an intimate one that creates numerous opportunities for bonding and that also holds the potential to be erotically charged. The combination devolves to the disadvantage of those who might need help that would take the form of conversation or touching. Able-bodied culture disdains any kind of mutually supportive activity in public toilets beyond small talk and, perhaps among women, the daring move of passing a toilet roll beneath the stall door. For the disabled, the accommodative technologies available are psychic compensations for their dependence; they render invisible, at least temporarily, the vulnerability and need for "help" that dare not speak its name. This happens coincident with desire by authorities to lower costs by dispensing with human care personnel—nurses, social workers, domestic caregivers, and so forth—whose role is not sufficiently valued to encourage their retention. Better to replace them with architectural supports or mechanical devices. Indeed, for some disability rights activists, the person who offers toilet assistance—a set of skills largely associated with hospital care or rehabilitation medicine—is regarded as a residual effect of the medicalization of the disabled body and thus anathema.

Ultimately, however, what is accomplished by assistive technologies in the public toilet is the loss of the exchange that otherwise might occur between human beings working with each other. Not only does the relationship between the disabled individual and the technology undergo a shift, but then so does the social connection. That is, the privileging of independence may underestimate the social and ethical values that accompany dependence: reciprocity, caring, and cooperation. This has profound implications not only for how we understand the disabled toilet user but also for how we understand the social networks in which both the disabled and able-bodied are embedded.

In a brilliant collection of essays on the theme of interdependence as a form of political organization, *Love's Labor*, philosopher Eva Feder Kittay offers a feminist critique of political philosopher John Rawls's influential work on the transformative possibilities of cooperation and mutual obligation in a democratic society.[18] Kittay argues that Rawls imagines a democratic ethos in which all citizens share in dividing labor equally among all "fully functional" members of the society but fails to recognize that within any social system there invariably will be individuals designated as dependents—children, the elderly, the temporarily infirm, and the chronically disabled—as well as those who are designated as caregivers, those on whom dependents rely for basic human needs. At the core of such a critique, Kittay argues, is the recognition that the lack of gender specificity in Rawls's model presumptively positions a heterosexually masculine and able-bodied figure at the core of democratic discourse, since women do the disproportionate amount of caregiving in our culture and have been, historically speaking, positioned more often than not as dependents themselves. "To model the representative party on a norm of a fully functioning person," Kittay writes, "is to skew the choices of principles in favor of those who can function independently and who are not responsible for assuming the care of those who cannot."[19]

Kittay's critique of Rawls's work has much to do with how we interpret the privilege given to the body that is "fully functioning" as well as the body that eschews dependence as a marker of weakness, vulnerability, and disempowerment. Dependence, especially as it relates to the care of the self, is problematic enough for both women and men; for each gender it raises some particular problems connected to gender. For men, it raises special problems for masculine self-realization, threatening emasculation. Manuals written by physicians and orthopedists for the care of disabled veterans during and after World War II, for example, traditionally emphasized recuperation as a restoration of male self-reliance and the maintenance of heterosexual masculinity.[20] And because of the freedom traditionally accorded to the male, heterosexual body in the public sphere, any body that requires some form of special accommodation or has particular needs is feminized as dependent. As a result, those who are recipients and/or facilitators of caregiving fight social stigmas on multiple fronts.

Although Kittay recognizes the potential for the dependent to challenge expectations of both a gendered and able-bodied public body, the greater hurdle to mutual respect and cooperation, it seems, is the way in which independence is valorized unproblematically as the ideological commitment to which the disability rights activist must pay unwavering loyalty. As Kittay writes,

> It is a source of great inspiration and insight in the disability community that independent living, as well as inclusion within one's community, should be the goal of education and habilitation of the disabled. But this ideal can also be a source of great disempowerment. . . . [The] focus on independence, and perhaps even on the goal of inclusion when inclusion is understood as the incorporation of the disabled into the "normal" life of the community, yields too much to a conception of the citizen as "independent and fully functioning."[21]

Kittay repudiates the value placed on the linkage between "independent and fully functioning" and what can be socially productive and socially valuable citizenship. In so doing, she is in effect contesting the classical demarcation of a gendered female private sphere versus the male public sphere. Privileged access to a public toilet fuses with privilege, male and able-bodied, more generally. The disabled toilet, then, is that which collapses the private—that which is gendered as female, domestic, and altogether dependent—into the public sphere, so that the disabled toilet becomes a type of *non*male space that is conspicuously neither publicly male nor privately female. Although the disabled toilet was perhaps envisioned as a material embodiment of equal access and democratic equality, it also expresses, by its nature, the masculine principle that slippage of gendered difference risks display of weakness and dependence.

One aspect of male performativity is the use of urinals, themselves totems of masculinity. But for certain types of disability, use of the urinal becomes problematic, and this poses gender-identity difficulties as disabled men turn to toilets as urination receptacles. Hence a manual for health professionals and rehabilitation specialists published in the early 1980s counsels on the need to teach disabled men to sit on the commode for "both bowel movements and urination," ensuring that it "avoids the difficulty that some handicapped men and boys have in aiming urine while standing so that it does not splash on the seat, floor, or clothing."[22] The task here, it seems, is how to give disabled men, and boys, the repertoire for performing bodily control, overcoming avoidance of a characteristically female toilet posture. Failing to build up men's inner reserves would efface their distinction with women, destabilizing the historical male compulsion to demarcate and maintain the feminine private sphere as entirely separate from the masculine public sphere.

For all newly disabled people there are challenges in acquiring and mastering toilet skills. Men bring special baggage of not only their own immediate pragmatic goals but also gender-based dependency anxieties. These come into play in dealing with, as an adult, the most intimate bodily functions: how to raise and lower oneself from a wheelchair to a toilet seat, for instance, or how to wipe oneself while balancing on grab bars—and maintaining dignity. On a routine basis, an ambulant person who is visually impaired must negotiate the space of a public toilet alone and deal with problems such as paper-towel holders placed well outside one's ordinary reach or toilet paper that has somehow gone missing from its usual location. There may be a slippery puddle, perhaps due to a missed maintenance call, that involves danger. In a biographical analysis of the life of Leonard Kriegel, a polio survivor who grew up in the 1940s, the historian Daniel Wilson has written, "What had changed [for Kriegel] was not the masculine values he embraced but the locale in which they operated, . . . not on the athletic field or on the field of battle but in the rehabilitation hospital and on the streets of Brooklyn as he struggled to rebuild his body and to confront the barriers of an unaccommodating society."[23]

Among the challenges to dealing with the anxieties at hand are the (literal) disabled signs that announce and confirm one's bodily separateness and, at times, regroup gender identities. Such a recalibration is evident in a photograph taken in May 2004 at the London Chelsea Flower Show, held annually since 1913 (now on the grounds of London's Royal Hospital)—the "world's greatest flower show," as proclaimed on the website (see fig. 8.1). In the photograph, one

Figure 8.1. Bathroom queue for ladies and disabled, London Chelsea Flower Show, May 2004. (Courtesy of the author)

can observe the banner reading "Ladies and Disabled Toilets." Event planners know that crowd management is an organic part of any toilet landscape, manicured or otherwise, but it is not always done right. In this case, the disabled share designation with the gender that is dependent and charged with caring for the dependent—children, the elderly, and the infirm. Under the prominent banner, toilets for disabled users are not only spatially linked to those of women but also are conceptually linked to that classic category of dependency, while the other population—men—maintain their unproblematic identity as separate creatures. Men share lines, let alone toilet stalls, with no one. Such a spatial polarization of facilities and choreographies becomes still more significant when one realizes the deliberate planning and attendant expenses involved in laying out the facilities, designating these as demarcated social spaces, and creating the actual signs.

This is a gendering of public toilets by design as well as the gendering of disabled toilets by default.

Writer Nancy Mairs observes in her memoir *Waist-High in the World* that "many of us with disabilities require some assistance, but with the right facilities we can maintain dignity for ourselves and those who care for us."[24] Such an insight suggests that Mairs understands that the technology of the disabled toilet does not necessarily eliminate the dependent relationship between the disabled person and his or her caregiver in the vaunted name of independence for its own sake. Instead, it transforms the relationship so that disabled people maintain some level of privacy and self-esteem while acknowledging the mutually beneficial dynamic inherent in the caregiver–cared-for relationship. Yet how does one provide the disabled toilet user with an environment in which he or she can use a facility that is both specially marked for accommodation and yet unmarked in terms of categorizations that stigmatize?

Section 504 of the Rehabilitation Act of 1973 and the Americans with Disabilities Act of 1990 remain cornerstones of federal legislation for people with disabilities and those who benefit from the same rights, including the people who give them care. It is worth recognizing, however, that federal guidelines for enforcing or invoking the ADA deliberately presume a disabled body that is unmarked or unaffected by differentials of gender, race, ethnicity, class, or sexuality, let alone bodily difference and normativity. As Jennifer Levi and Bennett Klein have shown, the ADA as well as the federal Rehabilitation Act explicitly exclude from protection any persons who claim to have been discriminated against on the basis of "transvestism," "transsexualism," or "gender identity disorders not resulting from physical impairments."[25] The strategy used by some trans people to attain an "official" medical diagnosis that confirms that their gender dysphoria constitutes a disability is not without controversy. Activists in the disability rights community disavow the medical model of disability, which claims that nonnormative bodies should be rehabilitated and "fixed" through technology to meet the expectations of a society deeply invested in the concept of physical and psychic wholeness.

The insistence on homogenizing the disabled as toilet users is reminiscent of a process that disability scholar Patrick White has described as the heterosexualizing of the blind, a process in which institutions and textbooks for the visually impaired have deliberately controlled and encouraged particular kinds of sexual intimacy and gender performance in situations where young blind people interact with one another.[26] Perhaps it is precisely this fear of sexual and gender nonconformity that surrounds ADA rules against protection for trans people and that, furthermore, has permitted the ADA to flatten out or erase outright the specific and often irreconcilable elements of one's private (and therefore gendered) bodily experience that constitute the terms of one's public (and therefore gendered) disabled identity. This misrecognition of the importance of gendered and sexual difference in the experiences of people with disabilities reinforces the impression that disabled citizens are interchangeable with one another. Disability is their assigned master status, with no other social or cultural identifications appropriate or necessary. Those identifications, in other words, are meaningless within the language of tolerance and formal equality, and insofar as they are recognized at all, they are gender-neutral. But as we have seen, the discourse around public toilets has never been gender- or sex-neutral but is inflected through and through with gendered prescriptions for autonomy and self-reliance, as well as, of course, with rights and privilege.

In Erving Goffman's 1977 essay "The Arrangement of the Sexes," he argued that toilets were spaces in which the socially constructed expectations of gender equality, the hard-won efforts of the sexual liberation movements of the 1960s and 1970s, did not hold sway. Segregation would remain. As Goffman asserts, "The *functioning* of sex-differentiated organs . . . [does not] *biologically* recommen[d] segregation; *that* arrangement is totally a cultural matter. . . . [Yet] toilet segregation is presented as a natural consequence of the difference between the sex-classes, when in fact it is rather a means of honoring, if not producing, this difference."[27] While Goffman may have regarded the gendered toilet as a retreat from or an alternative to the androgynous sexual politics of the mid-1970s, disability rights advocates in the main accepted gender segregation. For them, accessing

the women's or men's room was a way of *entering* the social, not re-
treating from it. Many were surely willing to take on and accept the
terms of a deliberately delineated sexual difference if it meant being
able to use a public toilet in the first place.

Yet even with the ADA's problematic and inherently homophobic
and transphobic language regarding gender and sexual difference, we
cannot in good conscience dismiss or dismantle its policies without
recognizing the enormous achievement that the bulk of its legislation represents for human rights in the United States and for disabled
people around the world. The work of contemporary transgender and
genderqueer activists who are taking on the public toilet as a site of
gender liberation borrows liberally from disability rights activism.[28]
Perhaps a more effective challenge to the ADA's policies regarding
trans and gender-nonnormative populations might be to ask, how
might a nongendered toilet policy transform the way that we talk
about gender and disability, recognizing that the use value of public
toilets is too often defined and delimited by the historical legacies of
heterosexual male privileges within public space? In other words, in
what ways can we harness the awesome power of the ADA to enact
effective spatial accommodations for people with disabilities while
also harnessing the power of gender-nonnormative critiques of the
public toilet?

There have indeed been some practical solutions offered up. Since
the 1960s, a growing group of architects, urban planners, and education specialists, as well as those in the fields of rehabilitation medicine
and palliative care, have been actively thinking about disability and
the built environment, including the status of public toilets.[29] One response, coming primarily from the practitioners of architecture and
planning, is the call for "universal design"—that is, the concept of
user-centered design for all ages and body types that eschews uniform
notions of a single, able-bodied, ergonomically normative subject, or
a single user. Single-occupancy unisex facilities provide an example.
Space otherwise given over to the separate men's and women's rooms
(and perhaps more space as well) is rearticulated as a series of small
walled-in enclosures, each large enough for wheelchair access and
that of a helper. Each has a sink as well as toilet. People who identify

as transgender would not have to make a choice, and those observing them would not be in a position to remark or reject. The stand-alone outdoor pay toilets, APTs, now common in Europe and appearing in some U.S. cities (see Braverman, chapter 4, in this volume), are a variant (fig. 8.2), with the New York versions, at least, of sufficient size to accommodate wheelchair users.

But there are drawbacks, in both ecological and financial costs. Use of urinals requires less space, less water, and less money for installation and maintenance than does toilet use. A plausible answer is to install urinals in a single unisex facility where men who wished to do so could use them for urination and both men and women could use toilet stalls as desired. Those with physical disabilities would use whatever appliance they preferred. This is a radical solution because it requires both men and women to share the intimate space of the restroom. It also, of course, eliminates the efficiency of restricting the disabled to one or two "special" spaces; every stall must be large. In the case of stand-alone pay toilets, provision of wheelchair access decreases the number of street locations large enough to accommodate them.

Numbers of architect-scholars treat all these concerns as a design as well as a political and moral issue. Some have recognized public bathrooms, as in the words of Raymond Lifchez and Barbara Winslow, "a focal point of the drive to create a barrier-free community."[30] Lifchez and Winslow were involved in the Center for Independent Living, founded in Berkeley, California, in 1972 by disabled students at the University of California, which became a paradigm for disabled community activists around the world and remains an active model for developing services and initiating antidiscrimination legislation. Indeed, the Center for Independent Living was among those Bay Area–based organizations that dispatched its activists to occupy the offices of the Department of Health, Education, and Welfare during the April 1977 protest actions. From the inception of organizations such as the Center of Independent Living, their goal has been to work simultaneously on developing legal strategies for protecting the disabled individual alongside those for transforming the built environment. Activists collaborate directly with urban planners, architects,

Figure 8.2a and b. Automatic pay toilet (APT),
Madison Square Park, New York, 2009.

technology specialists, and civic leaders to conceptualize urban land-scapes, with the toilet playing a central role.

The disabled toilet exemplifies the tension between autonomy and equality at the core of classical liberalism: How far do we go to rec-oncile the protection of individual freedoms with the need to satisfy the majority's needs and preferences? How much expense and incon-venience must the majority "suffer" to provide restrooms not only in adequate numbers for the disabled but also equipped for their diverse needs and respecting their sense of dignity, gender based and other-wise? Disabled people's use of public toilets focuses attention on the issue in particular ways. As historian Patrick Joyce has written, urban public space has long been the site of negotiation of this tension. It was in large cosmopolitan cities such as Manchester and Vienna dur-ing the eighteenth and nineteenth centuries where citizens learned how and where to exercise civic autonomy but also civic restraint in their direct physical engagement with urban thoroughfares, pub-lic squares, and municipal parks. Public spaces thus became political theaters where the productive strains of liberalism not only were ex-hibited but were materially manifested. While liberating individuals from the typical constraints of social life, the new urbanity restrained particular kinds of behaviors in an ongoing need to perpetually nego-tiate power relations that are both dynamic and unresolved. As Joyce writes, "If forms of power and human agency, and of bodily compe-tence and of knowledge, are carried in the material world, and in the *use* of objects," then one might characterize the history of modern lib-eralism as a narrative of a "strange and complex history of objects and material processes."[31]

Disabled people's use of the public toilet has indeed its own "strange and complex" history, part of which is the naturalization of the able and gendered body. Variations from this human model have, in countries such as the United States, enlisted an attitude of toler-ance but not a full embrace of the opportunities for celebrating and learning from difference. As always, more is eminently possible.

With a stance perhaps appropriate for a pioneer composer and performer, the British musician and disability activist Alan Hold-sworth (a.k.a. Johnny Crescendo) coined the slogan "Piss on Pity" in 1992.[32] It describes the attitude felt by many disability activists toward the patronizing compassion of the nondisabled, exemplified by the annual Labor Day telethon for the Muscular Dystrophy Association, hosted by Jerry Lewis, which promotes maudlin sentimentality instead of self-empowerment.[33] Combining the image of pissing as a natural biological function with the image of pissing as an aggressive act of individual and group disobedience, the "Piss on Pity" movement forces a recalibration of the stigma attached to disability, as propagated shamelessly by the medical-telethon-entertainment complex. Pity is not good enough, and indeed, it is destructive. Toilets remain contested terrain, with the question remaining of how, truly, to make all toilets, and all toilet users, equal before the law of the commode.

Rest Stop

Flirting with the Boundary

Exterior of Monica Bonvicini's "Don't miss a sec" mirrored public restroom, London, 2003–4.

FROM WITHIN THIS temporary public installation by the artist Monica Bonvicini, the one-way-mirrored walls appear transparent, keeping the user constantly aware of activity on the street. From the street the bathroom reflects its surroundings, protecting the privacy of insiders. The piece, one version of which was installed in front of Tate Britain, gains its frisson by raising even the suggestion of exposure.

Interior of Monica Bonvicini's "Don't miss a sec" mirrored public restroom, London, 2003–4.

NEXT STOP
20 PAGES

PART III

Building in the Future

9

The Restroom Revolution

Unisex Toilets and Campus Politics

Olga Gershenson

THE HISTORY OF the modern restroom has been a history of successive social groups proposing a right to access and a mode of toilet configuration fitting to their needs and desires. First were the women: we owe the term *potty politics* to the Ladies' Sanitary Association and similar women's organizations that put up a fight for the room of their own. Establishing the first women's lavatory in Victorian London took the persistence of the lone female member of the government and the advocacy of a famous vestryman, George Bernard Shaw, to overcome the resistance of the residents and the vestry. It was not until 1905, after five years of stalling, that the decision to build a women's bathroom was finally made.[1] A similar pattern, explained in part by Terry Kogan in chapter 7 in this volume, took place in the United States and other parts of the world.

In the U.S. case, racial minorities were next in line to take up the challenge. In the era of racial segregation in the United States, blacks and whites couldn't drink from the same fountain, let alone urinate into the same bowl, certainly across the South but to a degree in other parts of the country as well. Up until the 1950s and even the 1960s, locker rooms and bathrooms were still not integrated. For instance, when the Western Electric Company in Baltimore adopted a policy against segregation of public facilities, the union, consisting of white members, went on strike.[2] This was during World War II, and the strike had military consequences, leading the War Department to briefly take over the plant. Rather than continue the arrangement, after three months and further negotiation, resegregation was imposed.[3] As late as the 1970s, court battles over segregation of public facilities continued (e.g., *James v. Stockham,* 1977).

In line behind women and blacks were people with disabilities, who waged their own fight for full public participation via architectural modification of bathrooms, entryways, and more (see Serlin, chapter 8, in this volume). Their movement resulted in legislation that changed building codes and made the provision of an inclusive toilet a legal requirement.

Transgender and other gender-variant people have now joined the civil rights toilet queue, straining—with a mix of fear and indignation (and conciliation)—for admission to the public bathroom. Traditional sex-segregated public restrooms bring them routine risk of being insulted, mocked, attacked, and even arrested. As interim remedy, they hunt for bathrooms in which they feel stigma free and physically safe, timing their visits to avoid potentially conflict-ridden overlap with other users. The possible policy solutions to their problem vary from unisex or single-stall bathrooms to education campaigns that might, for example, cause others to be more open to the gender-nonconforming people in their midst. Whatever the best ultimate remedy, the demand for transgender toilet provision stirs public controversy, even among those sympathetic to the campaigns for civil rights of those who have come before.

Because transgender people do not conform to societal expectations of "male" and "female," their mere presence in a sex-segregated place raises anxieties about gender and sexuality.[4] Ever since psychoanalytic theory linked toilet training with sexuality, bathrooms and sex have been intrinsically connected in both public imagination and scholarly analysis. Patricia Cooper and Ruth Oldenziel, in their brilliant analysis of the cultural meanings of bathroom space, comment,

> The very creation of bathroom spaces, which are routinely separated by sex, reflect cultural beliefs about privacy and sexuality. Separating women's and men's toilet facilities prevents either sex from viewing, accidentally or otherwise, genitals of the other. The bathroom is a place where genitals may be touched but not primarily for the purpose of sexual stimulation. Western social norms also

dictate that sexual relations are not permitted here. The separation ignores same-sex sexuality. . . . So women's and men's bathrooms assume heterosexuality and the existence of only two sexes, permit genital touching, and reject overt sexual expression.[5]

These cultural meanings define the "discipline" (in Foucauldian terms) of the bathroom space, and hence its political meanings, reflecting "an unintentional cultural strategy for preserving existing social categories, the 'cherished classifications'"[6]—with the two-sex model key among them.

Restroom Revolution at UMass

These issues took concrete form in a political battle waged at the University of Massachusetts–Amherst (UMass) in 2001, when a student group called Restroom Revolution advocated for transgender-friendly unisex bathrooms on campus. The group's campaign, not atypical of similar initiatives at other U.S. institutions, resulted in an uproar in the campus media and a prolonged dispute with the administration. Conservative administrations resist changes proposed by more socially progressive student bodies. Other student groups resist as well—a pattern with precedent in regard to prior groups' push for civil rights. The opponents of unisex bathrooms expressed concerns over public morality, safety, cost, and regulatory intrusion—again a list with precedents in the rationales for opposing rights of prior groups when they made their demands.

Educational institutions are often held to be key mechanisms for realizing the American ethos of equal opportunity. It's no wonder, then, that colleges and universities have regularly found themselves at the forefront of activism and strife in the name of civil rights. Vietnam War protesters died for their cause at Kent State, the sexual revolution of the 1960s drew from a strong campus-based support network, and campuses were an important base for the civil rights movement, in both the North and the South. The women's movement was based in or at least abetted from campuses across the United States. Combine the widespread expectation that educational institutions will serve as models of secular social justice with their young students' passionate practice of identity politics, and it's no surprise that the fight for bathroom access plays out on campuses.

From a practical standpoint, the campus bathroom problem is doubly pressing for students whose workplaces and home spaces (dorms) so often rely on shared facilities. And such facilities are subject to regulation by both the state and university administrations, making them an appropriate target for demands to reform. Only a small percentage of campuses nationwide provide unisex bathrooms and gender-blind floors in residence halls, and their introduction is often divisive, as it was at UMass. The apprehensions reveal the persisting deep anxieties even in the socially progressive atmosphere of higher education and in places such as a liberal university in Massachusetts.

In the fall of 2001, several students gathered at the Stonewall Center, a lesbian, bisexual, gay, and transgender educational resource at UMass. They formed a special group to work on transgender issues on campus. As Mitch Boucher, one of the activists, explained to me, the issues included education about transgender people, especially for Resident Directors (RDs) and Resident Assistants (RAs), and the quality of life for transgender people on campus. It was at the meetings of this student group that the problem of bathroom use was raised for the first time. After sharing and reflecting on various risks and indignities suffered by transgendered students, the group's members reasoned that everyone, including gender-variant people, should be able to use campus facilities safely and comfortably.

The solution seemed clear: unisex bathrooms.

In December 2001, the group members sent their first proposal to the administrators responsible for student affairs and residence life on campus to establish several unisex or "gender-neutral" public bathrooms on campus. The students wrote,

> We are a contingent of students living in the residence halls of the University of Massachusetts who self-identify as transgender, transsexual, gender-queer, or something other than "man" or "woman." As gender variant people, we encounter discrimination in our daily

lives. The most pressing matter, however, is our use of the bathrooms in the residence halls in which we live. . . . We are often subjecting ourselves to severe discomfort, verbal and physical harassment, and a general fear of who we will encounter and what they will say or do based on their assumption of our identities.[7]

Further, the proposal made suggestions to designate at least one bathroom (including showers) per residence hall as unisex and to establish such bathrooms on every third floor of a building (in high-rise structures).

This original proposal had a didactic character: the Stonewall Center group wanted not only to bring about instrumental change by establishing unisex bathrooms but also to use it as opportunity for education and value change. They proposed to provide education on transgender issues "to all Residence Life staff." Moreover, they proposed to give information to students and parents not only about the location of these facilities but also about "why they exist." The subcommittee even tried to educate the administration by including with the proposal "a gender expression umbrella and a few of our personal stories."[8]

The administration was seemingly sympathetic: the vice chancellor of student affairs met with the subcommittee to discuss the proposal and, later, promised to establish two unisex bathrooms in the residence halls (in the Prince-Crampton and Wheeler dormitories). But nothing happened.

The next fall the students intensified efforts, issuing a call in the Stonewall newsletter *Queerie* for a meeting of all interested students. In response, on October 2, 2002, Restroom Revolution was formed. The twenty-three participants at this meeting represented a range of gender identities: transgender, transsexual, and gender-queer, as well as what the group called "allies"—straights who took up a just cause. The newly established group adopted the following mission:

> The mission of the Restroom Revolution is to advocate for safe, accessible restroom facilities for our campus community. . . . People who do not appear traditionally male or female risk harassment and violence in sex segregated facilities. The Restroom Revolution affirms the right of all people to have access to safe bathrooms.[9]

The first act of the group was to write another letter to the administration. The tone had changed. Instead of an emphasis on education, the group was trying to exert some political pressure: "It is the basic right of every human being to have access to safe, public restrooms, especially at a public university of which we are a part."[10] Besides the rhetoric of human rights, the group also put their struggle into the context of what was happening at other schools: "More and more, universities across the country are recognizing and responding to the needs of trans people by making gender-neutral bathrooms available to the campus community. We hope UMass, Amherst will follow their lead."[11] Indeed, among the schools that had unisex bathrooms at the time were Hampshire College, Amherst College, the University of New Hampshire at Durham, the University of Minnesota,[12] and the University of Chicago.[13] The majority of bathrooms at Reed College were gender-neutral.[14] Wesleyan University had in its residence halls a gender-blind floor with a unisex bathroom.[15]

The Restroom Revolution members didn't limit themselves to a letter. Mitch Boucher recalls, "We flooded the administration with emails and phone calls. We put posters in the bathrooms to raise awareness of the issue (see fig. 9.1). We set up a table at the Campus Center. We arranged for press coverage."

At the table at the Campus Center, Restroom Revolution distributed contact information for the university administration and a template to help in writing letters in support of the group. For those who didn't want to write letters, Restroom Revolution offered a petition of support they could sign. Hundreds signed. A student, allied with Restroom Revolution, produced and distributed a zine, *Lobbying for Lavatory Change* (see fig. 9.2). Finally, Restroom Revolution secured the support of the Student Government Association (SGA), the Graduate Student Senate, and the Graduate Employee Organization. The *Boston Globe* announced in its largely sympathetic article, "Transgender students at the University of Massachusetts at Amherst, unable to persuade administrators in the past year to create coed bathrooms, are shifting their strategy from private talks with school officials to a petition drive and mass mobilization."[16] A campus media debate ensued.

DO YOU KNOW THAT YOU ARE SITTING IN A SEAT OF PRIVILEGE?

The Mission of the Restroom Revolution is to advocate for safe, accessible restroom facilities for our campus community. Currently, many people do not have access to such facilities-- particularly those whose gender identity or appearance does not conform to societal expectations. People who do not appear traditionally male or female risk harassment and violence in sex segregated facilities. The Restroom Revolution affirms the right of all people to have access to safe bathrooms.

Stop by our table in the Campus Center Concourse to learn more about Restroom Revolution.

Figure 9.1. A poster distributed by Restroom Revolution and hung in the restrooms at UMass-Amherst in 2002. (Courtesy of the author)

Figure 9.2. Page from an independently produced zine lobbying in support of Restroom Revolution. (Courtesy of the author)

Media Positions

At first, campus media refused to take the bathroom issue seriously. A columnist in the official campus newspaper, the *Daily Collegian*, wrote, "When the Restroom Revolution at the University of Massachusetts started to hit the papers and the airwaves at the beginning of October, I think the initial reaction to it by most of the students on campus was predictable: they sighed, they laughed, and then they mouthed, 'You've got to be kidding me.'"[17] The *Collegian* quoted a member of the UMass Republican Club who opposed Restroom Revolution: "What the Republican Club sees it is a joke [*sic*]."[18] One of the responses on *Collegian* website ranted, "Get a life! There are much more critical issues to focus on. . . . In the real world, we seem to be worrying about an impending war in the middle east [*sic*], a real crappy economy, and a president who doesn't seem to know how to lead our country. Wake up! and take up a real cause."[19] The conservative monthly publication *Minuteman* ("Minutemen" is the university sports teams' name as well as a patriotic gloss) amplified the *Collegian*'s concern that Restroom Revolution was pursuing a silly cause. But the *Minuteman* went on to argue that Restroom Revolution's "joke" was not just harmless. Olaf Aprans worried that the activists were exaggerating concerns of no real significance in order to draw attention to the public display of morally repugnant behavior:

> The most probable motive for the Restroom Revolution is not the need or want of transgender bathrooms, it is the desire for attention. Transgender students have been using gender-specific bathrooms for years without any complaints. Why the sudden outcry for transgender bathrooms? The answer is easy, the activists behind this movement are using a petty issue like bathrooms as a medium to throw their lifestyles in the face of every-day students.[20]

In an article with a telling title, "The Politics of Pee," the *Minuteman* proclaimed, "gender-neutral bathrooms are neither an issue of safety nor comfort for transgender students; they are merely a means for homosexual activists to influence [the] campus with their immoral ideals and to break the traditional gender barriers that normal students hold."[21]

The vocal resistance of the on-campus press was echoed off-campus as other organizations took up the debate. Unsurprisingly, the Traditional Values Coalition, a nationwide organization of churches which routinely opposes campus-based efforts such as these, issued a condemnatory statement: "Individuals with mental problems should not be allowed to dictate social policies at a university nor [*sic*] in legislation that will normalize a mental disorder." It followed up with a special report on transgenderism, similarly censorious.[22] But resistance also came from those within the gay community. The online publication *Independent Gay Forum* ran a story entitled "Down the Drain" in which the author dismissed the petty concerns of pampered students and asked them to consider the truly meaningful negative implications for public relations between the queer community and the general public. In the author's words, "Certainly the transgendered, outside of the halls of ivy, face greater issues—like not being murdered. . . . Making straights use coed johns isn't going to improve matters in this regard."[23] The argument was that on-campus efforts draw broad negative attention to an already difficult fight for increased safety off-campus, where real danger lurks.

The discussion of unisex bathrooms brings responses of ridicule, indignation, and the conflation of gender nonnormativity with homosexuality. Evident are anxieties about sexuality and gender performativity and a primal fear of sexual mixing:

> There are only two things that make me a man and they are my X chromosome and my Y chromosome. . . . People have the right to feel that they should not be the gender that God gave them. . . . However, the fact that some people do not live in reality or that some wish reality were not true, does not entitle them to a special bathroom in a public university.[24]

As Cooper and Oldenziel explain in their research on the prospects for the integration of railways during World War II, "The specter of shared bathrooms not only touched people's feelings about privacy (feelings related in part to worries about one sex seeing the genitals of the other), but it also tapped deep fears about sexual mixing, transgressing social boundaries, and ending recognition of gender differences."[25]

The same fears were captured in an informal student survey conducted by and publicized in the *Minuteman*. The students were asked, "Do you feel that 'transgendered' students should have their own bathrooms, taking away a men's or women's room from buildings?"[26] The rhetorical impetus of the question is notable: unisex bathrooms accessible to all genders are turned into entirely segregated spaces from which presumably "normal" men and women are excluded. The answers were predictable: "No, that's ridiculous!" exclaimed a freshman. "No. If they are comfortable with who they are, they shouldn't be afraid to use the bathrooms and feel different from anyone else," said a junior.[27]

The anxieties surrounding sexuality lurked not only in the responses induced by the openly socially conservative *Minuteman* but also in the discussion on the *Collegian* website. In the responses dismissing the issue, the ridicule of gender variance became a major way of coping with these anxieties: "Its [*sic*] this simple: If you have a penis then use the men's room. If you do not have a penis then use the women's room."[28] Another respondent added, "If you want to be a woman, have some backbone and go get Mr. Happy chopped off. Until you feel strong enough to change yourself because of your beliefs, then don't you dare expect everyone else to change to cater to your needs."[29] Another joined in: "To ask for a separate bathroom because you dress as a member of the opposite sex, when in reality you are pretending to be a member of the opposite sex does not give you the freedom to have a bathroom designated specifically for you. . . . You are male or female, end of story."[30]

Claiming naturalized sex differences is a problematic and much contested, albeit familiar, argument. But even if there are "natural" biological differences between men and women, this argument is beside the point, as "the 'natural' is not, and never has been for human beings, the sole determinant of social possibility."[31] And it was this "social possibility" that Restroom Revolution's opponents feared. A columnist in the *Collegian* summarized the attitude of Restroom Revolution's opponents, who were reluctant to face the more complicated picture of the social world:

> Proponents of the [Restroom] Revolution argue that sex is a social construct, that it's something that society put in place to restrict our freedoms. They couldn't be more right. The thing is, it's something that's been in place since the beginning of time. Challenge the existence of male versus female? Next perhaps we'll argue what color the sky is.[32]

Making claims based on "natural" differences is, of course, a recurrent strategy over history, including the efforts that went into defeating the Equal Rights Amendment for women; antiratificationists wanted to preserve, as Jane De Hart explains, "classification by sex, the definition of social role by gender." For them, she continues (and as further examined by Mary Anne Case in chapter 10 in this volume), "gender . . . was sacred. It was a given: a biologically, physically, spiritually defined thing, an unambiguous, clear, definite division of humanity into two."[33] It was the fear of losing the sacred division of gender into two that propelled opposition to Restroom Revolution.

Safety First, but Safety for What and Safety for Whom?

The issue of safety was tightly woven into the issue of the preservation of gender binaries or the ability to relax them. Advocates of unisex bathrooms had argued for unisex facilities to enhance the safety of gender-variant people. In Boucher's *Collegian* editorial, he argued that establishing unisex bathrooms "is a safety issue, as well as a workplace issue for RAs, grad student employees, faculty and staff whose gender expression or identity does not conform to cultural norms."[34] The administration, on the other hand, took a position that it was a matter of discomfort, not necessarily safety. Boucher paraphrased a member of the administration who downgraded the safety argument to one of discomfort: "Well, you guys can use the bathrooms, you are just uncomfortable." This rhetoric sets the mild "discomfort" of gender-variant people against the "real" issue of women's safety.

According to the reports in both the *Collegian* and the *Minuteman*, the issue of safety became a stumbling block during the SGA meeting that discussed the resolution of support for Restroom Revolution. The *Collegian* quoted student senator Matt Progen, an opponent of change: "This has nothing to do with sex or gender; it has to do with making some people feel comfortable and the comfort of some is not as important as the comfort and safety of others."[35] Another senator quoted in the same article, Dave Falvey, was clearer about who stood to lose if gender-variant people gained unisex bathrooms: "I think it's a bad idea and I'm concerned with the safety of women on this campus. . . . I feel that it's taking women's safety away. Basically I feel that it's a very big safety concern and putting women in a more vulnerable position."[36]

The *Minuteman* quoted Falvey in greater detail, revealing that the anxiety incited by the proposed acceptance of gender nonconformity was rooted in fears of male-on-female violence:

> What would prevent . . . a drunk male at three a.m. in the morning . . . from entering this "unisex" bathroom and attacking or harassing a female? . . . It's bad enough that these types of attacks are occurring *with* sex segregation in place. Can you imagine how many opportunities for this type of incident would arise if males were *allowed* to be in the shower stall next to their female peers?[37]

Another SGA senator wrote a letter to the *Collegian* amplifying the paranoia Falvey expressed, warning women that in unisex bathrooms they would be surrounded by voyeurs and pranksters:

> I have several friends who can see over the stalls. As a female, could you ever feel comfortable knowing somebody can walk in next to you and lean over at any time? . . . We are a campus that has several incidents of reported rape each year. Worse off [*sic*], more common examples of abuse would probably be simple and immature in nature, like stealing a towel.[38]

In the humor section of the *Minuteman*, in which the paper's "Jackass of the Month" title was awarded to Restroom Revolution advocate Ed Kammerer, a similar argument appeared:

> Kammerer gave very little thought to the safety of women in bathrooms. The first reported rape of the year happened in the bathroom of Chadbourne dormitory. Common sense dictates that if there are frequent encounters of men and women in bathrooms (possibly when drunk at 2:30 a.m.) there may be an increase in situations.[39]

Accepting a gender-nonconformity approach threatens to unleash male violence and mischief that's already just barely contained.

Of course, women's safety is an important, even vital, concern on college campuses and beyond. However, the anti–Restroom Revolution rhetoric employs particular traditional attitudes toward masculinity and femininity to frame how safety can be achieved: weak females require protection from violent male sexuality. This ideology relies on the production and maintenance of clear performances of and boundaries between two genders. Conveniently, these boundaries would not only protect women's private "safe" place but also keep them in their place, disciplined and properly sorted.

Administrative Response

With all the letter writing, petitioning, and news coverage, how did the administration respond? Slowly, quietly, and with heavy reliance on dry state-based regulatory language. Although some students were willing to fight the powers that be, the administration made reference to those (still higher) powers in support of the status quo. In an interview with me, the vice chancellor of student affairs at the time pointed first to Massachusetts architectural code: "In the case of dormitories, there is a regulation [as to] how many fixtures you have to have in a building, how many for men, how many for women. It also says that the bathrooms have to be designated by gender." What about Restroom Revolution's rhetoric of human rights and the widely accepted framing of the struggle as a political fight? The vice chancellor's attitude was patronizing, advising caution lest the group might have been unaware that they were inviting negative attention: "My first response to Restroom Revolution is, you are looking for problems, you are setting up a target. You are bringing attention to yourself, to your political statements."

The administration did conduct a "building summary," collecting data about bathrooms on campus. The survey showed an insufficient number of bathrooms—which was offered as evidence of the impossibility of converting gendered bathrooms to gender-neutral bathrooms. Under the administration's strict reading of the regulatory code, a bathroom without a gender mark was no bathroom at all. Thus, changing the designation of a bathroom from male or female to unisex would lower the already insufficient number of bathrooms on campus. But from a regulatory standpoint, the worst that could happen if several bathrooms were designated unisex and therefore uncountable as bathrooms under the code would be an inspection failure. In that case, the university could either undertake to build more bathrooms or appeal to the Board of State Examiners of Plumbers and Gas Fitters in Boston. "Have you checked this possibility?" I asked the administrators. They recoiled in response: "No, no, we know it's virtually impossible to get a permit from the Board."

However, Joseph Peluso of the Board of State Examiners of Plumbers and Gas Fitters, with whom I spoke on the phone, assured me that "the regulation cannot cover all the situations" and that the Board encourages organizations to "file for a variance if they have an extraordinary situation." The university could have appealed to the Board on the grounds that it had an extraordinary situation, one covered by state disability laws. According to Lisa Mottet of the Transgender Civil Rights Project, in Massachusetts transsexual people are indeed protected under state disability laws.[40] Therefore, there is no legal difference between handicap-accessible and unisex bathrooms. But the administration was not motivated to explore this direction.

The vice chancellor also employed a numbers game, assuming that the population desiring unisex accommodation at the school was obvious by sight: "Some of the challenges are to assess the needs. I saw two [transgender] students—I know that one individual already graduated—and even though I received many letters of support, I took a position to meet the needs as they are demonstrated. I don't see the need right now to increase the program." It appears that the administration did not view the "many letters of support" for Restroom Revolution or the hundreds of signatures on the petition as evidence of "need" within the student body.

In the end, the group was never taken seriously because its members' demands were interpreted as an expression not of a real need but of a desire to be more "comfortable," and for a handful of people at most. These factors—a narrow understanding of "the needs," the interpretation of Restroom Revolution as ploy, and the pressure of negative publicity—defined the administration's strategy of doing the minimum. It appeared sympathetic to a sensitive issue by changing the signage on two single-user restrooms of the many hundreds on the entire campus while neatly sidestepping a broader reevaluation of gendered restroom provision.

The controversy over unisex bathrooms on this contemporary college campus in a liberal state echoes historical controversies over the creation of public restrooms for women and the destruction of racially segregated restrooms. It reveals deep cultural anxieties about the consequences of a slowly eroding gender binary. As with past exclusionary practices, the mere fact that the cultural practice is widespread or typical (such as war or slavery) does not make it just or desirable.[41] Philosopher Richard Wasserstrom asks "what a good society can and should make of these [gender] differences."[42] His answer is unequivocal: "eradication of all sex-role differences."[43] Why? Because maintaining a system of gender categorization inevitably leads to sexist attitudes and practices, of "taking sex into account in a certain way, in the context of a specific set of institutional arrangements and a specific ideology which together create and maintain a *system* of unjust institutions and unwarranted beliefs and attitudes."[44]

Sex-segregated bathrooms, Wasserstrom concludes, are just "one small part of that scheme of sex-role differentiation which uses the mystery of sexual anatomy, among other things, to maintain the primacy of heterosexual sexual attraction central to that version of the patriarchal system of power relationships we have today."[45] The same patriarchal system that envisions sex as a crucial binary category insists on the sexual segregation of bathrooms.

Feminist philosopher Louise Antony continues the argument for the eradication of gender. Echoing Wasserstrom, she argues, "Gender, whatever its etiology, is the raw material of sexism. . . . We should strive for a society in which biological sex has as little systematic social significance as eye color."[46] This does not mean, however, that she advocates the obliteration of difference. "Under current, gendered arrangements, difference among human beings is 'packaged' into two salient, mutually exclusive categories. . . . The point of abolishing gender categories, and with them gender norms, is to eliminate such homogenizing forces."[47] So the result would be more difference, not less.

Rest Stop

Thai Students Get Transsexual Toilet

WITH SPACIOUS, TREE-LINED grounds and slightly threadbare classrooms, there is nothing obviously unusual about the Kampang Secondary School. It is situated in Thailand's impoverished northeast, and most of the pupils are the children of farmers. Every morning at 8:00 a.m. they all gather outside to sing the national anthem and watch the flag being raised.

Then they have a chance to use the toilets, before heading off for the first classes of the day. Kampang is proud of its toilets. Spotless and surrounded by flowering tropical plants, they have won national awards for cleanliness.

But there is something else about them too. Between the girls' toilet and the boys', there is one signposted with a half-man, half-woman figure in blue and red. This is the transsexual toilet, and outside, in front of the mirrors, some decidedly girly-looking teenage boys preen their hair and apply face cream.

The head teacher, Sitisak Sumontha, estimates that in any year between 10 and 20 percent of his boys consider themselves to be transgender—boys who would rather be girls. "They used to be teased every time they used the boys' toilets," he said, "so they started using the girls' toilets instead. But that made the girls feel uncomfortable. It made these boys unhappy and started to affect their work."

So the school offered to build the transgender boys their own facility. Triwate Phamanee is a slightly built thirteen-year-old who is adamant

Transsexual bathroom sign
at school in Thailand.

that he will one day change his gender. "We're not boys," he told me, "so we don't want to use the boys' toilet—we want them to know we are transsexuals." Vichai Saengsakul, fifteen, agreed. "People need to know that being a transsexual is not a joke," he said. "It's the way we want to live our lives. That's why we're grateful for what the school has done."

Tolerance, said transgender rights activist Suttirat Simsiriwong, is not the same thing as acceptance. Despite the high profile of transsexuals in Thailand, they complain that they are still stereotyped—they can find work easily enough as entertainers, in the beauty industry, in the media, or as prostitutes, but it is much harder to become a transgender lawyer or investment banker.

Text used with permission from BBC News; original article by Jonathan Head.

10

Why Not Abolish Laws of Urinary Segregation?

Mary Anne Case

PUBLIC TOILETS ARE among the very few sex-segregated spaces remaining in our culture, and the laws that govern them are among the
very few in the United States still to be sex respecting, meaning that
they still distinguish on their face between males and females. It is
this, rather than the experience of having to wait on one too many a
long line for the ladies' room, that led me to put questions of sex discrimination in the provision of public toilets on my scholarly agenda.
In examining the history of the development of the constitutional
law of sex equality, I was struck by the vehemence with which Phyllis
Schlafly and other opponents of the proposed Equal Rights Amendment (ERA) in the 1970s insisted that its passage would mean a mandatory end to restrooms segregated by sex. Leaflets urging voters to
reject the ERA even claimed that it was "also known as the Common
Toilet Law."[1]

Although the ERA did not pass, other prominent items in the ERA
opponents' parade of horribles, such as an end to legal prohibitions
on same-sex marriage and on women in combat, no longer seem far-
fetched. For the most part, however, public toilets remain sex segregated. Even in public spaces, such as restaurants, where two single-
occupancy, self-enclosed toilet facilities are all that is provided to
customers, signs designate one "Stallions" and the other "Fillies," one
"Pointers" and the other "Setters," or, more prosaically, one "Ladies"
and the other "Gents." Usually this is a product of the requirements of
the law, as innumerable state and local ordinances specify that there
be "separate free toilets for males and females, properly identified, on
the premises."[2]

To be sure, some efforts to integrate toilets by sex have made head-lines over the years. College student Wendy Shalit catapulted herself to national attention and a book contract by editorializing in 1995 against the vote of her Williams College classmates to make their dor-mitory bathrooms co-ed.[3] A few years later, the use of a unisex toilet as a prominent plot device in *Ally McBeal,* a TV show set in a ficti-tious Boston law firm, led a few actual firms to experiment with uni-sex toilets of their own.[4] More recently, transgender rights advocates have gained some traction on college campuses with calls for gen-der-neutral restrooms. Yet, as Olga Gershenson details in chapter 9 of this volume, even when advocates ask only that a few, not that all, public toilets on a given campus be open on a gender-neutral basis, their request stirs up the sort of opposition that would delight Phyllis Schlafly. In 2004, for example, after administrators at the University of Chicago acceded to a request by the Coalition for a Queer Safe Campus that about a dozen of the hundreds of bathrooms on cam-pus be made gender-neutral and that future construction on campus make provisions for gender-neutral restrooms, Rush Limbaugh was one of several nationally prominent conservative commentators to express outrage. "Feminists support equality," Limbaugh said. "Look what has to happen to institutions in order for these people to secure equality. You have to weaken the institution, in this case male and fe-male bathrooms."[5]

Contrary to Schlafly's earlier prediction, however, the call for a nonsexist restroom is not typically a call for mandatory unisex toi-lets (see, for example, Greed, chapter 6, in this volume). The call, instead, is more often for "potty parity," a term of art for more eq-uitable provision of separate toilet facilities for men and women. In response, states and municipalities throughout the United States have put into effect dozens of potty parity laws since 1987, when the Cali-fornia state legislature passed a bill that State Senator Diane Watson informally dubbed "the 'parity in potties' measure." This bill had been introduced by Watson's colleague Senator Art Torres after his wife, Yolanda Nava, reported being stuck for over half an hour at the the-ater behind a restroom line of more than fifty women, some of whom had finally invaded a nearly empty men's room.[6] Perhaps because he and his wife compared notes, Senator Torres understood quite clearly what so many regulators and users of public toilets still do not to this day: "Restrooms are the same size in most facilities, but urinals in men's rooms take less space than" stalls.[7] Urinals lead restrooms equal in square footage to offer more excreting opportunities to men than to women. When such features as fainting couches, full-length mirrors, and vanities are added—as they sometimes are—to women's but not to men's rooms, the ratio of excreting opportunities given equal square footage gets even worse for women.

The fact that "public life [is] subject [to] laws of urinary segregation," to repeat the words of Jacques Lacan (mentioned in chapter 2), often keeps these inequalities from view.[8] We cannot know how the other half lives or what is behind the door to the restroom we are forbidden from entering. But, notwithstanding the failure of the ERA, we do tend to assume in the modern United States some measure of sex equality. Thus, too many people casually assume that behind the restroom door they cannot enter are facilities equal to those available to their own sex. For example, male students at the University of Virginia Law School were surprised to learn, in the mid-1990s, that their female counterparts had full-length mirrors available in the restroom; these men then promptly requested mirrors of their own, so they, too, could preen before a job interview. Men make demands.

The comparative paucity of excreting opportunities behind the door marked "Ladies" might become evident from observation of the comparatively longer lines often outside that door. Instead, both men and women tend to attribute those lines to the fact that women take longer once inside, perhaps simply because they spend so much more time on primping and powder-room gossip. Studies carried out by researchers do offer some statistics in support of the assumption that women take longer, with one of the most widely quoted finding that women take an average of seventy-nine seconds, men forty-five, in the restroom.[9] Many of the available studies do not distinguish, however, between time spent waiting on line, time at a stall or urinal, and time at a sink or mirror.

There are many reasons, of course, why women might indeed take longer than men actually using the toilet. As Judge Ilana Rovner put it in her opinion dissenting from her colleague Judge Richard Posner's holding that an electric company's failure to provide "civilized bathroom facilities" for its only female lineman was not sexual harassment, "The fact is, biology has given men less to do in the restroom and made it much easier for them to do it."[10] Culture works against women as much as nature does. Whereas pantyhose slow women down, for example, the zipper front and center on a typical pair of pants only facilitates male urination. As Harvey Molotch has ironically observed, "If women truly want to relieve themselves as efficiently as men, they can take some initiative. Options do exist short of biological alteration,"[11] among them changes in clothing styles.

In addition to nature and culture, the role of the law in creating those long lines should not be underestimated. At the time Senator Torres introduced his potty parity bill in the late 1980s, "the three major model plumbing codes (BOCA, Southern Standard, and Uniform) in the United States specif[ied] minimum elimination fixtures (water closets and urinals) for men's restrooms that [we]re often greater than the number for women's restrooms . . . depending on the type of facility, the specification formula used, [and so on]."[12] These codes began with the nineteenth-century premise that women were less likely to be out and about in public than men were, a premise that could become something of a self-fulfilling prophecy. Thus, Clara Greed describes the urban women of today doing what blacks were forced to do in the Jim Crow South—carefully planning their day to take account of the very few places legally available for them to excrete.[13]

As litigation against Jim Crow in the first half of the twentieth century demonstrated, a demand for facilities that are "separate but equal" is one possible response to blatant inequality. Inevitably, though, practical questions as to exactly what is to be equalized and how plague any regime of separate but equal. Just as segregated railroads had difficulty determining in advance exactly how many dining-car seats to set aside for black and white patrons, so, for example, an ice rink that hosts a hockey game on one day and a figure-skating competition on the next may face widely varying ratios of male to female patrons, and therefore widely varying demand for toilets. If equal square footage is indeed too empty and formal a measure of equality, should equal facilities or equal excreting opportunities be the goal? Or should the goal be to equalize waiting time? Should one then take into account that women may take longer? What allowance should be made for the fact that more young boys tend to accompany their mothers into the women's room than girls accompany their fathers into the men's room?

Torres's bill sought to remedy inequity by requiring plumbing codes to take full account of the number of women likely to use a facility. Other early potty parity laws defined parity as a one-to-one ratio of excreting opportunities for men and women. Interestingly, long history provides more support than the recent past for such a ratio. The famous Whittington's Longhouse, a public toilet built in medieval London with funds specifically bequeathed for the purpose by Lord Mayor Dick Whittington and kept in operation on the banks of the Thames until the seventeenth century, had 128 seats—64 each for men and for women.[14]

Increasingly, the trend in potty parity has been to require the construction of more excreting opportunities for women than for men. The Texas potty parity law, for example, introduced in the uproar following Denise Wells's 1990 arrest at a concert for entering and using a men's room after finding thirty women ahead of her in line for the women's room, mandated twice as many women's as men's toilets in new or renovated public spaces.[15] Even so, potty parity laws, however well intentioned, still leave unremedied many existing inequalities in favor of men. And on the rare occasions when men perceive themselves to be the victims of inequality, they are less patient and long-suffering than women have been. When the renovation of Chicago's Soldier Field in accordance with local potty parity law led to longer wait times for men, who constituted more than two-thirds of the audience at Bears games, male protests led to the conversion of five women's rooms to men's rooms, and the reestablishment of wait times for women that were on average twice as long as those for men.[16]

Why isn't the simplest solution, then, to end sex segregation in public toilets? From the moment I first began seriously to consider the question of sex equality in toilets, I assumed it would be. More

specifically, I thought I would be recommending as a model something like the typical airplane bathroom, a facility used by members
of both sexes, one at a time, in complete privacy. This would be consistent with the approach I have taken in my law journal writings
to other situations in which sex distinctions have been abolished in
law—instead of assuming that what was previously available to men
is appropriate for everyone, I have urged consideration of sameness
around a feminine standard.[17]

Unfortunately, the typical pattern when sex distinctions are abolished is that women are offered what had previously been available to
men. For example, in recent decades, women have been encouraged to
enter the work force in large numbers. A far smaller number of men,
with very little societal encouragement, have become the primary
caregivers of their children. Once on the job, women all too often find
everything from the uniforms to the performance standards to the
working hours tailored for the men for whom the jobs were once reserved. The temptation to shoehorn women into an environment built
to suit men plagues public toilet design as well. As the introduction to
this volume notes, a number of more or less complicated devices are
marketed to facilitate a woman's using a urinal like a man. An example
is the P-mate (fig. 10.1), "a moulded paper funnel" which, "when positioned securely under the crotch, and with underwear pushed to the
side, . . . directs urine away from the body to a suitable place, such as a
toilet, a container or a conveniently located tree."[18]

If devices designed to encourage a woman to urinate more like a
man ever were to catch on, they might themselves generate cultural
anxiety, as the toilet scene in the film *The Full Monty* indicates. In it,
unemployed steelworkers spy on their wives, who have taken over the
local Workingman's Club for an evening of entertainment by male
strippers. The steelworkers come upon women occupying the men's
room, cheering on one of their number as she hikes up her skirt
and directs a stream of her urine into a urinal. Already threatened
in their masculinity, the men conclude, "when women start pissing
like us, that's it, we're finished, extincto. . . . They're turning into us.
A few years and men won't exist, except in zoos. . . . I mean we're not
needed no more, obsolete, dinosaurs, yesterday's news."

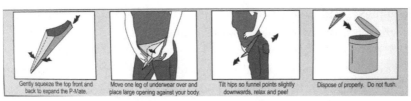

Gently squeeze the top front and back to expand the P-Mate.   Move one leg of underwear over and place large opening against your body.   Tilt hips so funnel points slightly downwards, relax and pee!   Dispose of properly. Do not flush.

Figure 10.1. P-mate portable urinal for women.

Devices such as the P-mate, Shewee, She-Pee, She-inal, I-Pee, Brief Reliefs, and Safety Urinette have yet to gain widespread acceptance, perhaps because they often involve complicated paraphernalia. Technologically simpler efforts to encourage men to adopt urination methods associated in the Western world[19] with women have also been seen as threats to masculinity. Feminists in Germany have been urging men to accustom themselves to urinate while sitting on a toilet seat by posting signs in restrooms with the imperative *"Hier wird sitzend gepinkelt"* (Here one pees sitting down) and by explaining that such a practice would be more sanitary and create less work for those responsible for cleaning toilets, who are most often women. While some men have taken pride in accommodating this demand, others have vehemently resisted, going on talk shows, publishing editorials and cartoons, and forming Facebook groups of *"Stehpinkler"* (Those who pee standing up). So vehement was the resistance that academic Klaus Schwerma, a proponent of *Sitzpinkeln,* could write an entire critical book entitled *Stehpinkeln—Die Letzte Bastion der Maennlichkeit?* (Peeing Standing Up—The Last Bastion of Masculinity?).[20]

There might be some benefit to men in adopting more of the excreting practices now associated with women, however. If the model of the airplane toilet, a model much closer to the toilet stall in a typical women's room than to the urinal in a typical men's room, were to become the universal norm, ending sex segregation in the toilets need not mean a loss of privacy for women. It could instead offer increased privacy to men, something that could appeal at least to those men who suffer from pee-shy, a pathological inability to urinate easily when at risk of being observed, as at the urinals in a public toilet. (It is worth asking why, except in rare cases such as Japan Air's specially equipped "Ladies' Elegance Rooms," airlines have not attempted sex segregation of their toilets. Even Japan Air acknowledges that men, too, can use the Elegance Rooms because "there is probably no way we could enforce absolute discrimination on an aircraft."[21] Could it be that, especially in the days when norms of air travel were developing, men were the overwhelming majority of airplane passengers, such that reserving even one, let alone an equal number, of scarce airplane toilets for women would leave male passengers waiting on long lines, something that, unlike women, men would not quietly tolerate?)

Basic queuing theory confirms that making fully enclosed single-user facilities available to either sex on demand, as airplane toilets are, would cut down on overall waiting times and promote the most efficient use of available toilet facilities. To some extent, a drive toward efficiency is indeed motivating the construction of such toilets. When fully enclosed single stalls are increased in size beyond the typical airplane size to the dimensions of a toilet accessible by the handicapped, the potential efficiencies increase exponentially, as do the number of disparate constituencies whose needs are met. Consider the increasingly popular creation of relatively spacious, single-stall, fully enclosed public toilets labeled for use as Family/Handicapped/Unisex. Such toilets relieve a number of anxious dilemmas, such as that of a mother sending her young son alone into the men's room without her, the adult son waiting outside the door of the women's room for his Alzheimer's-afflicted mother to emerge, and the wheelchair-bound husband left to navigate the handicapped stall in the men's room without the help of his wife. When such toilets also include a sink within their fully enclosed space, as they typically do, they facilitate the ritual ablutions that observant Muslims find more difficult to perform in stalls removed from access to running water.

Such toilets also relieve the anxious dilemmas of the transgendered or those who, whether or not intentionally, read as gender liminal or otherwise outside a clear gender binary. After all, walking into a toilet segregated by sex requires that each of us in effect self-segregate by hanging a gendered sign on ourselves—and I do mean gendered even more than sexed, given that the signs we are asked to choose between are typically pictograms of a stick figure with a skirt and one without (men and women identified by gendered dress) and rarely, if ever, ♂ and ♀ (males and females identified by their genitalia). Some individuals have profound objections to hanging any one of these binary signs on themselves; others would be challenged if they made the choice they wished. Of those who would be challenged no matter which of the two sex-segregated restrooms they used, some identify as transsexual or transgendered, but others, including, for example, some butch women, emphatically do not. Without a unisex restroom, all who do not clearly read as male or female are faced with the prospect of challenge, even of assault or arrest, no matter which door they walk through, with the result that many report that they do their best to forgo use of public toilets altogether.

For those, like me, whose vision of sex equality includes an end to what the U.S. Supreme Court has called "fixed notions concerning the roles and abilities of males and females,"[22] there are therefore feminist as well as practical efficiency payoffs from the increasing popularity of the Family/Handicapped/Unisex restroom. Individuals will not be forced to conform to any standard of what it is appropriate for a man or for a woman to look like in order safely to enter a public restroom. Other forms of gender nonconformity will be made easier as well. Fathers and other male caregivers will find it much easier to be out and about in public with young children if they have reliable access to a restroom to which they can accompany those children in comfort and privacy. Perhaps this will encourage them to do so more often. Exactly this mixing up of sex roles in society at large was what ERA opponents most feared.

Notwithstanding my hope that the time for integrating toilets by sex may have come, a majority of men and women in the United States—not only retrograde opponents of women's equality but many who themselves identify as feminist—would still oppose abolishing the laws of urinary segregation. There is apparently a vast distance between what I would prefer and what many other women prefer. Not only do many women object to sharing a restroom with men, whom they perceive as less tidy, as well as potentially more threatening, many women also value the women's room as a site of female sociability.

Let me respond to each of these grounds for continued segregation in turn, beginning with the notion that sex-segregated toilets keep women safer from attack. My response here begins with the anecdotal observation that an awful lot of male-on-female crime already takes place in the supposedly safe space of the women's room. My perusal of sources ranging from newspapers to law reporters indicates that robbery, assault, molestation, rape, and even murder are not infrequently perpetrated by men who have followed or lain in wait for women and girls in the toilet. Even male-on-male crime can occasionally take place in a women's public toilet.[23] Nor are women safe in the segregated toilet from male bad behavior that may stop short of crime. For example, plaintiff Mechelle Vinson, whose case before the U.S. Supreme Court firmly established hostile-environment sexual harassment as an actionable form of sex discrimination in employment, testified that, among many other bad acts, her male supervisor "followed her into the women's restroom when she went there alone" and exposed himself to her.[24] Similar allegations appear in a number of other sex-harassment cases.

When I sought to quantify the amount of male-on-female crime that takes place in the women's room, I found myself stymied by the absence of crime statistics with this information. (Let me digress for a moment to complain that the lack of readily available reliable data plagues almost every aspect of the study of public toilets. It can be almost as hard to find reliable information as it can be to find a public toilet when you need it.) In the unfortunate absence of data, I turn to one particularly well known and horrifying incident to help make my case that what sex segregation provides to women may not be safety but instead the illusion of safety, an illusion that can itself prove deadly. In 1997, seven-year-old Sherrice Iverson was murdered in a stall of the Las Vegas casino women's room into which she had fled to escape from eighteen-year-old Jeremy Strohmeyer, who had been chasing her.[25] She must have been thinking, "I'll be safe here. The sign on the door means this is someplace he can't follow me." But, of course, he could and did; and there, in a locked stall, he molested and killed her. Although the sign on the door could deter some men with criminal intentions from entering a women's room and could draw immediate regulatory attention to others when they try to enter, the potential expected presence of both sexes in an integrated restroom could also on occasion act as a deterrent, by decreasing the likelihood a perpetrator will be alone with his intended victim and increasing the chances a bystander able and willing to offer aid will be present.

When women describe the women's room as a safe space, they generally have in mind much more than physical safety, however. They see it as a place to escape from a browbeating boss or importunate suitor, a place where they can cry without being seen and gossip with one another without being overheard by any man, a place where they can literally and figuratively let their hair down. The notion of women's restrooms as a haven may carry over from attitudes toward the far greater number of separate public spaces reserved in earlier centuries for women only, as Terry Kogan describes them in chapter 7 in this volume. It is interesting to observe that, at least for some, the colored restroom could serve much the same function in the Jim Crow South. Thus, John Howard Griffin, a white journalist who darkened his skin so as to report his experiences living as a black man in the segregated South of the late 1950s, repeatedly describes the colored restroom both as a place of sociability with others "black like me" and as a refuge from the insults of the white world. Griffin writes that when he "could stomach no more of this degradation," he "entered one of the cubicles [of the men's room] and locked the door": "For a time, I was safe. . . . In medieval times, men sought sanctuary in churches. Nowadays, for a nickel, I could find sanctuary in a colored rest room."[26]

This is only one of many reasons I am inclined to question those who rest their claim that separate but equal—an unacceptable solution for race segregation in toilets—might work for sex segregation on the assumption that sex-segregated toilets play a completely "different role in our culture than did racially segregated ones."[27] The philosopher Richard Wasserstrom, for example, in an important early article comparing race and sex discrimination, insists that, whereas racially segregated toilets were connected to an "ideology of racial taint" which held that blacks were "dirty and impure" and should not

be permitted to "contaminate bathrooms used by whites," the ideology behind sexually segregated bathrooms contains "no notion of the
possibility of contamination or even directly of inferiority or superiority" but merely a need to maintain "that same sense of mystery . . .
about the other sex's sexuality which is fostered by the general prohibition on public nudity."[28]

It may be true that the fear of contamination flows in both directions for sexually segregated toilets in a way it doesn't for racially
segregated ones—women, after all, cite the mess men make in the
toilet as a reason not to integrate—but to deny any notion of contamination behind sex segregation of the toilets is to blink reality, as
I learned while observing the integration of women into the hith-
erto all-male Virginia Military Institute (VMI) in the late 1990s.
In litigation opposing the admission of women to VMI, the school
made much of the educational benefits afforded by "total lack of
privacy," with male cadets under constant observation even while
in "gang bathrooms."[29] Admitting women, the school successfully
convinced a lower-court judge, would have one of two unacceptable consequences: either the women, too, would "lack . . . privacy,"
thereby "destroy[ing] any sense of decency that still pervades the
relationship between the sexes,"[30] or "adaptations would have to be
made, in order to provide for individual privacy," thereby destroying
equality, transparency, and the VMI honor code, which depended,
according to Judge Jackson Kiser, on "the principle that everyone
is constantly subject to scrutiny by everyone else."[31] Although noting that the educational system in Plato's *Republic* featured both
sexes exercising together in the nude,[32] Supreme Court Justice Ruth
Bader Ginsburg acknowledged for a Court majority that "admitting
women to VMI would undoubtedly require alterations necessary to
afford members of each sex privacy from the other sex."[33] VMI officials went Justice Ginsburg one better in giving the women whom
the Supreme Court required be admitted some privacy even from
their own sex. In fact, even before the advent of women on campus, the school's lack of privacy was less than total: the men's toilets at VMI did not feature just one large trough but separate stalls
with waist-high wooden partitions, stalls which lacked doors mainly

because the school got tired of replacing those broken off their hinges by rowdy cadets. But the women not only got toilet stalls with doors that closed; they got individual curtained shower stalls rather than open, communal showers like the men. When I asked why this difference, a male cadet muttered something to me about "health reasons," while his commanding officer amplified with reference to "blood-borne diseases."[34] The indefinable expressions of distaste in the voices and faces of both these men confirmed for me that that women to them, like blacks to their Jim Crow predecessors, were "dirty and impure" and that, if anything, segregation of the toilets, perhaps by preserving precisely that mystery about the bodies of the opposite sex on which Wasserstrom focuses, fosters the conviction that sharing space with women threatens with the possibility of contamination.

Moreover, women who seek refuge in the women's room, as John Howard Griffin did in the colored washroom, do so in part because the men's room in some environments can function as something like the executive washroom, a point reinforced in those academic institutions which, years after the admission of women, still had a sign reading simply "Faculty" on the door of a men's room. A woman can escape her boss in the office women's room only if the bosses are men. The flip side of this safe space for female subordinates is a safe space for male bosses, free from the intrusion of women seeking professional advancement. Popular culture reinforces this, as in film after film, the uppity professional woman gets her comeuppance when she is stopped at the door of the men's room. "I'd love it if you weren't here," says the newspaper publisher played by Jason Robards to Glenn Close in the role of his high-level subordinate in the film *The Paper*. She has followed him through an open doorway into a large men's room at a black-tie function to protest his refusal to renegotiate her contract. As the other men turn and stare, she is forced to retreat in ignominy. Separate public toilets are one of the last remnants of the segregated life of separate spheres for men and women in this country, now that the rules of etiquette no longer demand that the women leave the men to their brandy and cigars after dinner in polite company. Although the spaces can be made separate but equal, then and now, the access to power offered by an all-male and an all-female space continue to differ enormously.

A few years ago at a conference, I presented a paper on the cultural uses made of sex-segregated restrooms with the title "On Not Having the Opportunity to Introduce Myself to John Kerry in the Men's Room." I got the title from 2004 Democratic presidential candidate John Kerry saying to *Daily Show* host Jon Stewart that what most surprised him in his presidential campaign was the number of people who tried to introduce themselves to him in the men's room, an opportunity I will never have. More generally, given the repeated insistence by men that conversation in the restroom is taboo for them, I find it noteworthy how much networking does seem to go on in the men's room. Several junior male lawyers, for example, have told me of getting assigned to major cases as a result of restroom conversations with senior male partners. And one senior male litigation partner at the major New York firm at which I used to work was notorious for beginning business conversations with male subordinates with the invitation, "Come pee with me." It is worth noting that the Bohemian Grove, the ultimate men's power club, whose membership in recent years has included U.S. presidents, Cabinet officials, and other male power brokers, defines the ability of members to urinate freely together on the trees in the Grove as both the core of their bonding experience and the principal reason why female members would be unthinkable.[35] Maybe the reason why some male journalists complained that women had an unfair advantage covering Hillary Clinton and some male comics write routines suggesting that women use the restroom as a power center is because each is projecting from his men's-room experiences.

It is clear that one answer to the question "What if anything important might we lose if the laws of urinary segregation were to be abolished?" is "the opportunity to be alone with one's own," with one's own now defined by sex in a way that it once also was by race and class. (There was, after all, a time when "Ladies" and "Gentlemen" were elite subsets of and not mere synonyms for "Women" and "Men.") But is the opportunity to be alone with one's own, when one's own are defined by sex, a cost or a benefit of the laws of urinary segregation? Many people, among them many women, clearly continue to see it as a benefit. I am, I must admit, even after careful consideration of the competing arguments, more inclined to see it as a cost. In this, somewhat perversely, I may see eye to eye with Phyllis Schlafly—we each suspect that the achievement of equal rights for women may entail an end to sex segregation in the public toilet.

Rest Stop

Menstrual Dilemma

TAMPON PACKAGING ADVERTISES "flushable" qualities. Bathroom signage urges users not to flush. What is the right thing to do? Flushing tampons, pads, applicators, and anything other than toilet paper will threaten the functioning of many plumbing systems. Pads block plumbing immediately; just three tampons over time can back up a house.

But when we trash rather than flush, are bloody used feminine products biologically hazardous waste, threatening the health of others, especially those who clean out the frequently unlined disposal bins? And what about menstruation in the unisex bathroom? Will there be greater pressure to flush to prevent the possibility of discovery (by sight or smell) by men who may harbor greater disgust than women about the remnants of menstruation?

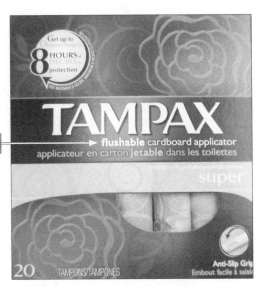

"Flushable" instructions on a box of applicator tampons.

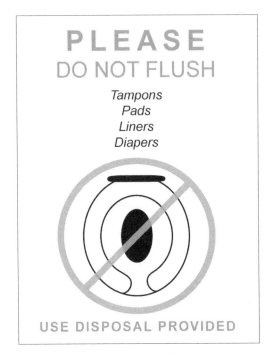

No-flush sign from a restaurant bathroom.

11

Entangled with a User

Inside Bathrooms with Alexander Kira and Peter Greenaway

Barbara Penner

IN THE PAST decade, it has become common for scholars to study toilets through the analytical lens of "discipline": following Foucault, they are concerned with the question of how such spaces articulate and uphold various forms of social difference.[1] This chapter, however, considers discipline in another, more literal sense and explores instead what toilets reveal about the disciplinary limits of architecture itself. As the historian Hayden White noted in *Tropics of Discourse*, "Every discipline [is] constituted by what it *forbids* its practitioners to do," although in this case it is more apt to say "by what it forbids its practitioners to speak about" or even how it allows them to speak.[2] Toilets threaten to contaminate the purity of architectural discourse—its discipline—by contaminating the divide between high and low, matter and spirit, temple and outhouse, on which it still implicitly depends. In one recent polemic, the coeditor of the *New Criterion*, Roger Kimball, dismissed toilets as an unthinkable subject of serious academic enquiry: "But really: a book about the ideology of public toilets? Has it come to that?"[3]

It has. As this chapter establishes, toilets have not been invisible in art and design discourse—far from it—but they have been spoken about in very particular ways in order to contain their taboo aspects. The first section of this chapter explores the ways in which toilets have been "cleansed" within architecture, specifically through the modernist language of formalism. In the second and third parts, this chapter turns to two projects which challenge this cleansing process and reveal the nature of what others avoid: architect Alexander Kira's groundbreaking attempts between the 1950s and the 1970s to investigate toilets from the perspective of use and filmmaker Peter Greenaway's dazzling 1985 film *26 Bathrooms,* which exposes the usual restrictions of expert discourse and portrays, in contrast, a reinsertion of bathrooms into the routines and rhythms of daily life. Through these two works, this chapter seeks to identify what can be gained by discussing toilets within academe. What is gained by such disciplinary acts of "contamination"?

Striking Lines

Although a discussion of strategies of spatial purification could easily begin at almost any point in the history of civilization, the cleansing of architectural discourse is most overtly linked to the cleansing drive of modernism.[4] To start with a discussion of modernism is appropriate in more ways than one, as we immediately find a toilet appliance at its foundations: Marcel Duchamp's *Fountain.* Its history is so well known that it needs only a brief rehearsal here. In 1917 in New York, Duchamp submitted a standard-model urinal to the Society of Independent Artists for an exhibition. His only alterations were to rotate the object ninety degrees, to name it *Fountain,* and to sign it, "R. Mutt 1917." Rejected from the show itself, the piece came to be known through the Dada journal the *Blind Man,* in which, beside Alfred Stieglitz's iconic and artfully lit photo (made at Duchamp's request),[5] an editorial made the claim that continues to reverberate today: "Whether Mr. Mutt with his own hands made the fountain or not has no importance. He chose it. He took an ordinary article of life, placed it so that its useful significance disappeared under the new title and point of view—created a new thought for that object."[6]

Although this gesture and proclamation have undoubtedly been crucial for the rise of modern conceptual art, this particular scandal is also revealing about the visual language of modernism. Even though Duchamp later claimed that he never selected his "readymade" objects on the grounds of beauty (the choice of readymades, he said, was based on "visual indifference"),[7] almost immediately an aesthetic argument in favor of the urinal was articulated. The ceramicist Beatrice Wood in her memoirs recalled the defense of the urinal by Duchamp's patron and friend Walter Arensberg in the face of a fellow judge's repulsion: "A lovely form has been revealed, freed from its functional purpose, therefore a man clearly has made an aesthetic contribution." Significantly, he continued, "If you can look at this entry objectively, you will see that it has *striking, sweeping lines.*"[8]

In the decades to come, it became increasingly acceptable in visual and architectural representations to treat toilets as sculptural objects and to find beauty and value in their sweeping lines. In 1924, the Modern Movement's most important polemicist and practitioner, Le Corbusier, reproduced an image of a mass-produced bidet in his journal, *L'Esprit Nouveau,* at the head of an article, floating above the caption "Other Icons: The Museums," that questioned the types of objects typically privileged in museum. He proposed instead "a true museum, one which contained everything," including "a bathroom with its enameled bath, its china bidet, its wash-basin, and its glittering taps of copper and nickel."[9] And just one year later, in 1925, one of the most influential modernist photographers, Edward Weston, celebrated the toilet's form in *Excusado,* a series of platinum prints taken over a two-week period, featuring the unadorned bowl, base, and bolts of his Mexican toilet (*excusado*). In an oft-quoted entry from his *Daybook,* Weston confessed that he found the experience revelatory, enthusing, "I have been photographing our toilet, that glossy enameled receptacle of extraordinary beauty. . . . My excitement was absolute aesthetic response to form."[10] (This photo is also significant because it shows that, at this level, the editors of the *New Criterion* were happy to talk about toilets: indeed, Hilton Kramer, the journal's founder, praised *Excusado* as "an essay in pure form.")[11]

Standardized, white, and pristine, toilets appealed not only to modernist artists and architects such as Le Corbusier and Weston but also to historians of modernism such as Sigfried Giedion. In *Mechanization Takes Command* (1948), Giedion represented the compact American bathroom as a triumph of progress and chronicled, with distinct empathy, its struggle to gain its "natural form" in the face of Victorian "weakness for adornment." Freed at last from "grotesque" ornamentation, in Giedion's eyes, the functional, manly simplicity of twentieth-century bathroom equipment perfectly exemplified modernist aesthetics.[12] It is with good reason, then, that artist Margaret Morgan has described the toilet as the "grand signifier of twentieth-century Modernism," an ur symbol of values such as cleanliness and of the machine aesthetic itself.[13]

As a result, the toilet was often the first line of attack in the 1960s and 1970s by critics eager to challenge modernist dogma. Critic Charles Jencks observed that "the lesson of the toilet bowl" was a favorite among architectural semioticians, who noted that newly installed toilets were being used in the south of Italy to wash grapes, thus exposing the arbitrariness of the link between form and function.[14] And the visionary engineer Buckminster Fuller, in an example prominently cited by Reyner Banham, used toilets to demonstrate the limits of Modern Movement architects' embrace of technological culture. He complained that the modernist enthusiasm for sanitary fittings almost never led to a practical concern for their design, their use, or the environmental problems that they represented; modern architects, he said, never "went back of the wall-surface to look at the plumbing. . . . They never enquired into the overall problem of sanitary fittings themselves."[15] While Fuller was bemused by this failure, his comments touched on why they held back. The sort of investigation of walls, cavities, sanitary fittings, and in situ use that he proposed would have fundamentally compromised architecture, a discipline that continued to define itself in important ways in opposition to engineering—and certainly to plumbing.

As Beatriz Colomina remarks, for instance, it would be a mistake to interpret Le Corbusier's reproduction of an image of a bidet in *L'Esprit Nouveau* as being a Duchampian gesture. In her view, Le Corbusier absolutely adhered to the traditional distinction between "the object of use and the art object"; for him, she suggests, the bidet was simply part of "folklore," the product of anonymous processes of production, whose proper place was a "museum of decorative arts," not an art gallery. Colomina's argument convincingly explains why Le

Corbusier was content to reproduce the bidet *as it was* rather than altering its design or, say, turning it upside down and rendering it unusable, that small but radical gesture that allowed Duchamp to give his urinal a "new thought" and to claim it for art. Le Corbusier's writings affirmed his belief that true artistic creation occurred only when an individual artist inscribed his creative will onto the materials around him for all time. This, then, was the difference between the Sistine Chapel and filing cabinets: "First the Sistine Chapel, that is, works where passion is inscribed. Then, machines for sitting, for classifying, for illuminating, *machine-types,* problems of purification and cleanliness." The latter machines, including the bidet, were of a "second-order."[16]

It is evident that to Le Corbusier art was not everyday and functional; it was lasting and transcendent. Or to map this onto the historian Nikolaus Pevsner's later, seminal definition of architecture, it was not a bicycle shed; it was Lincoln Cathedral.[17] Reyner Banham noted that Pevsner's formulation turned on aesthetic intentions, effectively creating an opposition between *modo architectorum* and "numerous other modes of designing buildings," for instance, of the sort that would produce a bicycle shed or a public toilet.[18] This focus on *modo architectorum* lies at the heart of architectural practice and architectural history and, as Banham well knew, excludes most ordinary spaces or mass-produced designs from consideration. As a result, architecture conventionally only interests itself in a very small proportion of the built environment: mostly the products of named architects.[19] And with a few notable exceptions such as Giedion (who pointedly subtitled *Mechanization Takes Command* a work of "anonymous history"), works by named architects remain the proper subject of most architectural and design discourse even today. As Rem Koolhaas has deadpanned, "God is dead, the author is dead, history is dead, only the architect is left standing."[20]

There is, of course, another reason for the debased status of spaces such as toilets, highlighted elsewhere by Koolhaas: the association of plumbing with femininity. In *Delirious New York,* Koolhaas memorably describes the 1931 Beaux-Arts ball in Manhattan, where famed architects came dressed as their skyscrapers to perform "The Skyline of New York": among others, William van Alen dressed up as the Chrysler Building, and Leonard Schultze donned the Waldorf-Astoria. Surveying the stage full of skyscraper-architects, Koolhaas notes the presence of only one woman: Miss Edna Cowan, the "Basin Girl," who carried a sink with hot and cold taps, the plumbing to the male architects' skyscrapers. Koolhaas writes, "An apparition straight from the men's subconscious, she stands there on the stage to symbolize the entrails of architecture."[21] Miss Edna Cowan represented a basin, a common receptacle that, in the formalist economy of the Beaux-Arts ball, was the negative of the creative, upwardly thrusting forms of the male architects.[22]

The gendered symbolism of this tableau is no accident. Adrian Forty has argued that, within architecture, ideal conceptions of "form" are almost always implicitly or explicitly based on a male body.[23] When not associated with decorative appliqué of the designed surface, women are, by contrast, associated with the interior fittings of buildings—the insides and services (or "entrails," as Koolhaas bluntly puts it). Although this servicing role may be presented as "natural" to women, the feminine is not here projected onto natural forms but rather merges with the mechanical and the technological. As the bearer of plumbing, the "Basin Girl" joins a well-established lineage of modernist figures, including the robot-woman in Fritz Lang's *Metropolis* or the "Mechanical Bride" discussed by Marshall McLuhan, which fuse the female body, sex, and technology and which work to contain the dirty underbelly of modernism.[24]

"Basin Girl" reminds us once again that modernist championing of the bathroom and the toilet was at once much more circumscribed and particular than it initially appears. In the works of writers such as Giedion, they became emblems of rational, mechanical production processes, even though, as Ellen Lupton and J. Abbot Miller have argued, they could equally and more aptly have become emblems of modern acts of consumption and waste.[25] But production was stressed, consumption suppressed, no doubt because mentioning use would have required invoking the individual, gendered, experiencing body—the body that eats, digests, and urinates or defecates. Toilets were icons of functionalism, but their function remained unmentionable. Visually, this suppression manifested itself in the way that the toilet was represented in modernist texts, free-floating, disconnected from pipes or from any person that might initiate use. This was one of *Fountain*'s basic lessons: in order to appreciate the urinal's "sweeping lines," its form actually had to be freed from rather than follow its function.[26] Toilets could be icons of functionalism only if their function was not mentioned. This suppression was vital to sustaining the technological illusions of formalism and architectural purity itself.

Alexander Kira's The Bathroom

Historical surveys aside, it is striking how little study and debate toilet design has generated within the fields of architecture or planning, given that there are few spaces that potentially have more impact on sanitation, mobility, and civility—all significant concerns of twentieth-century urban discourse. The sole exception to this rule is Kira's one-hundred-thousand-dollar Cornell-based study, conducted between 1958 and 1965. Kira's purpose was to completely rethink bathroom design. Published in 1966, the report detailing the study's results, *The Bathroom*, remains almost unique: it attempts to consider and to accommodate all aspects of human lavatory requirements, physical and psychological, in an objective way.[27]

The Bathroom's most important section, part 2, established a basic set of design criteria for bathroom equipment with the aim of producing safer and more appropriate fittings.[28] Even Buckminster Fuller would have approved of the comprehensiveness of the proposed redesigns, not to mention Kira's own criticisms of those willing to make do with existing products and conventions; he was particularly damning about the poor design of toilets themselves, noting that the insistence on providing users with a seat and, worse, making it too high, rather than encouraging squatting, was at the root of many modern health problems.[29] As Kira wanted to accommodate the optimal number of users most comfortably, he drew on anthropometric data about average bodily dimensions, taken mainly from military studies.[30] Despite his use of such standardized data, Kira was very aware of the different needs of various groups of users, from children to the aged, and always studied women and men performing personal hygiene activities separately. He was also aware of the different contexts and settings for toilet use.

These concerns came to the fore in Kira's revised and expanded edition of *The Bathroom,* published in 1976. In it, Kira included a new part on public hygiene facilities, which he believed were used differently—mainly, he said, they were used more perfunctorily because people were rushed and had less privacy—and, hence, had different requirements from residential ones.[31] As well as surveying the history and the social and psychological aspects of public facilities, he included criteria for designing and planning them, taking into account physiological needs and accounts of daily use. In his suggestions for women's facilities, for instance, Kira began by pointing out that women typically do indeed hover over toilets to urinate, due to fears about dirt and disease. In itself, this observation was not new: the period between the 1950s and 1970s had seen the introduction of female urinals by major toilet manufacturers, whose product names paid open tribute to these hygienic concerns ("Sanistand" in the case of American Standard and "Hygeia" in the case of Kohler). But, Kira remarked, these models did not sufficiently take into account the "management problems" posed by women's clothing or the postural limits imposed by undergarments, perhaps why they were never popular. In contrast, Kira proposed a modified women's urinal that, by replacing the conventional seat with angled thigh-high supports, held users in an optimal hovering position with a minimum of disrobing and of bodily contact (fig. 11.1).[32] However radical the design appeared, it was both practical—it was modular to allow prefabrication—and common sense: it showed a detailed understanding of the way in which women really use public toilets and respected cultural values such as privacy. Recognizing that women have to disrobe to urinate much more substantially than men do, Kira also insisted that his redesigned women's urinals, unlike men's, be in stalls.

As this example shows, Kira generally refused to treat toilet spaces as purely formal or scientific problems. In his design criteria, he

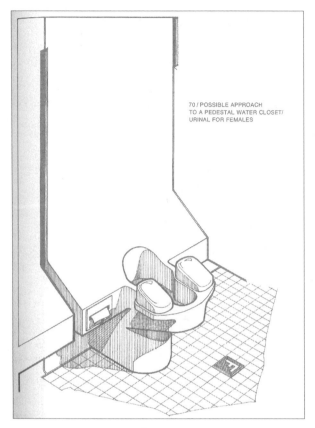

70 / POSSIBLE APPROACH
TO A PEDESTAL WATER CLOSET/
URINAL FOR FEMALES

Figure 11.1. Modified urinal. (Alexander Kira, *The Bath-room*, rev. ed. [New York: Viking, 1976])

aimed to give cultural and behavioral considerations equal weight as he gave to anatomical and ergonomic data, arguing that it is counterproductive to ignore them, as toilet "producers" and "purchasers" (by which he meant architects and builders) regularly did.[33] He himself described his criteria as "a compromise between realities as they exist [i.e., anatomy] and as they are defined with our time/place setting [i.e., cultural framework]."[34] With his deeply pragmatic sensibility, Kira made other compromises, too. He saw, for instance, that "taboos and guilts" around his subject were the greatest obstacle to

Figure 11.2. Getting into the bathtub. (Alexander Kira, *The Bathroom*, rev. ed. [New York: Viking, 1976])

The Bathroom's success, so that, although never aestheticizing toilets, he adopted other strategies to contain their psychosexual associations and the anxieties they might provoke in readers.[35] He invented a scientific terminology to describe bathroom activities: toilets became "hygiene facilities"; bathing became "body cleansing"; and urinating or defecating became the rather sinister "elimination."[36] When he did depict human subjects in photographs, he strove to produce a similarly neutral air, not only putting his models in bathing suits but also decorously blanking out their faces with strips of paper (fig. 11.2).

The effect of these strips is at once startling and surreal, and distinctly at odds with the frank tone of much of Kira's writing, not to mention the aims of his book. Kira obliquely acknowledged and excused the gesture when, in the revised edition, he noted that his new illustrations were "more natural than seemed possible a decade ago."[37] The dramatic change between the first and second editions is most apparent in the photographs accompanying the later work: the bathing suits are gone, along with the modesty shields, and models of both sexes are shown naked. In one study, a female model even demonstrates what happens when urinating from a standing or bent position, against a grid that measures the distance of her urine's trajectory (fig. 11.3). One writer cites the publication of this image as proof that, by 1976, "the social revolution had done its work";[38] and certainly books such as the two pathbreaking Kinsey reports (to which *The Bathroom* is sometimes compared), the sexual response studies of Masters and Johnson, and Alex Comfort's illustrated *The Joy of Sex* (1972) must have prepared the ground for such explicit images. However, unlike *The Joy of Sex*'s naturalistic freehand drawings, *The Bathroom*'s photographic studies still claimed scientific legitimacy through the backdrop of the grid, which, as Peter Greenaway notes, strongly evokes Eadweard Muybridge's human motion studies.[39]

The book's objective visual language was further carried through into its charts, diagrams, plans, elevations, sections, and line drawings. These remind us that the book's core project—the adoption of Kira's design criteria—depended in part on its being professionally legible. In this sense, it is not surprising to find that Kira's technical drawings

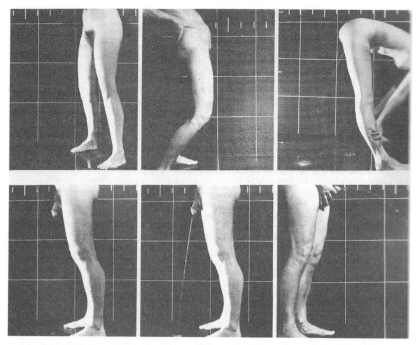

Figure 11.3. "Comparative male and female urination postures." (Alexander Kira, *The Bathroom,* rev. ed. [New York: Viking, 1976])

and studies broadly conform to the conventions set out in architectural "bibles," handbooks such as *Architects' Data* or *Architectural Graphic Standards,* and the standardized manner in which they depict interiors, including residential and public toilet facilities, and normative human dimensions, reach, and movement.[40] In one example from *Architects' Data,* diagrams of a residential water closet and sink show gendered but generic human figures going about daily activities and establish space requirements and clearances to accommodate them. (Conforming to the international standard, measurements here are in millimeters; Kira provided his in both metric and imperial units.) However much Kira elaborated on this approach, his work still owed much to this technocratic graphic tradition that emphasized data and suppressed expressive techniques such as shading or mimetic likeness in favor of line.[41]

There were probably ideological as well as practical and professional reasons why this was so. For instance, Kira would have looked favorably on the aesthetic neutrality of the graphic standards, as he was highly suspicious of fashion and believed the quest for stylish bathrooms led to unsuitable designs. To his readers, he suggested, "The 'best,' and real luxury, could also be something that functions superbly instead of something ordinary fashioned from pretty and costly materials."[42] For this same reason, apart from potted plants, he rarely specified materials or included ornaments in his views. And, of course, the graphics standards' matter-of-fact approach to the human body and its needs would further have defused the taboo nature of his study. Yet one consequence of using such highly schematic visual studies was that the nonprofessional may well have found them baffling or downright bizarre. One wonders, for instance, what the average reader made of Kira's many plan views of humans seated on toilets (fig. 11.4). As in *Architects' Data*, these

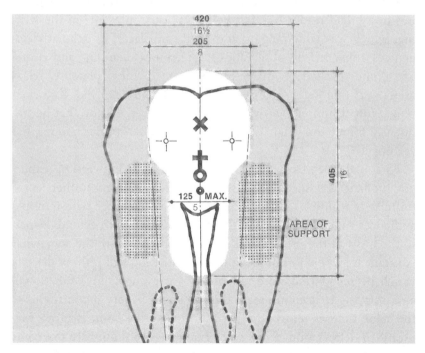

Figure 11.4. "Plan view of ladies' urinal." (Alexander Kira, *The Bathroom*, rev. ed. [New York: Viking, 1976)

plans delineated the necessary dimensions and clearances for bathroom equipment; unlike the bird's-eye view of the plans in *Architects' Data,* however, Kira's were done from below, in order to show both male and female genitalia, the void of the toilet hole, and the placement of necessary supports for the body. Such images pushed the graphic standards' omniscient and supposedly neutral viewpoint to their limits.

There is no doubting the sincerity of Kira's desire to initiate wider public discussions surrounding bathroom space and to counter the obfuscation surrounding toilet use and habits. Even though *The Bathroom* began life as an academic study, its revised edition obviously targeted a mass audience and came with an endorsement by consumer champion Ralph Nader, then at the height of his fame. Unlike the original publication, the revised edition also featured many crowd-pleasing photos of the latest bathroom fixtures and fittings, Kira's reservations about fashion aside. And in distinct contrast to practical manuals such as *Architects' Data,* Kira's text reached out to a broad readership. His voice was very "present" in the writing, which was wide-ranging and chatty in places, its scholarly references enlivened by observations, anecdotes, memoirs, and risqué jokes, and dwelled on many of the same themes (pleasure, relaxation, sex) that Peter Greenaway's *26 Bathrooms* later did. Kira even poked fun at himself, wryly noting in his revised edition that he frequently found the original edition of his book "buried among the sex novels."[43]

Kira might well have described the incongruity between his expert graphic language and his accessible text as being yet another "compromise of realities," a multipronged strategy to engage with multiple audiences simultaneously. But the problem seemed to extend deeper, to the heart of the project itself. Kira found himself in the somewhat contradictory position of using functionalist tools to rethink designs which he recognized were not purely (or even mainly) functional, and of trying to accommodate existing social beliefs and anxieties at the same time as reforming them. To make this point is in no way meant to detract from Kira's work but rather to illuminate the complexities one faces when one moves away, as Kira sought to do, from a strictly formal approach. How does one redesign a taboo? The question remains pertinent, as toilet taboos have proved remarkably resilient in the face of change—the final frontier of taboos, now that sex is no longer unspeakable in public.

Peter Greenaway's 26 Bathrooms

The tensions between subjective and objective modes and the gaps and silences that surround bathroom use are knowingly probed by Peter Greenaway. He produced his 1985 film *Inside Rooms—26 Bathrooms* for the UK television station Channel 4. *26 Bathrooms* presents twenty-six bathrooms in twenty-six minutes. Greenaway arranges the bathroom scenes in alphabetical order, so that they correspond to letters: "B is for Bath," "C is for Cleanliness," and so on. Greenaway deliberately moves his audience through a range of bathroom environments—monastic, luxurious, bleak, high-tech, Victorian, public— and uses these episodes to explore ideas about intimacy, cleanliness, and pleasure from a personal viewpoint. People deliver their thoughts while they are actually in the bathroom, or they are simply shown using it: bathing, shaving, washing the dog, reading, breakfasting, relaxing, chatting, exercising, singing, or watering the plants. We see one man having a smoke and reading a newspaper while sitting on the toilet ("Q is for Quiet Smoke") and a couple using a Jacuzzi ("J is for Jacuzzi"). They are young, old, middle-aged, white, black, posh, not posh, men, and women. Like *The Bathroom*, the film overall displays a great deal of humanity, if that is not too unfashionable a word, and humor of a gentle kind, rather than the blacker sort which is often Greenaway's trademark. Interestingly, Greenaway does not indulge in toilet humor and resists the scatological, if not the sexual.

Yet even when the film hits a truthful chord, it is anything but realistic. It is scripted, rigidly structured, and visually seductive. Accompanied by Michael Nyman's beguiling score, it features long and lingering shots of rippling waters, reflective surfaces, and ceramic fixtures that are no less gorgeous than Weston's photographic still lifes. The many dramatically lit tableaux of flowers, plants, shells, bottles, and bowls of fruit inevitably bring to mind Dutch interiors and still

Figure 11.5. Still from Peter Greenaway's *26 Bathrooms*. (Courtesy of Peter Greenaway)

lifes (fig. 11.5). Several scenes are simply surreal: in "H is for Hi-Tec [*sic*]," a woman in a leotard and dangling plastic earrings eats a cucumber while leaning against a wall of green studded rubber, an open umbrella reflected in the mirror (fig. 11.6).[44] The letters of the alphabet that structure the film often have whimsical referents: "A is for A Bathroom" pointedly refers to the indefinite article "A" instead of the space itself, and "S is for the Samuel Beckett Memorial Bathroom" refers to one man's associations of bathing with wartime privation. (This character returns in the concluding scene, "Z is for a Zoological Note," in which he shares his suspicion that "bathing is not terribly good for you" and leaves us in the realm of the melancholic paranoid.) As intended, these devices work to reveal the illusionism and partiality of the documentary genre. Greenaway calls *26 Bathrooms* an "artificial documentary"; by "artificial," he does not mean that it is "fake" or a spoof but that it exposes the methods and conventions of documentary filmmaking as it goes along.[45]

To this end, one of Greenaway's most effective gestures is to inter-
cut his characters' meditations with that of an expert on bathrooms
("X is for an Expert on Bathrooms"), a design professional who lec-
tures viewers on the history of plumbing and the theories behind
toilet design—miasmic theory and hydraulics—and mentions key
names in the field, including that of Kira (fig. 11.7). Indeed, the Expert is a partial homage to Kira, but the portrait is not an exact one:
Greenaway's aim is not to make fun of *The Bathroom*, which, even in
2009, he invokes with amused respect.[46] (Kira is also dedicated his
own letter, K, and images from the revised edition of *The Bathroom*
are shown in the film.) Rather, the Expert, one of the only characters
whose voice recurs throughout the film, allows Greenaway to skewer
the artifice, and occasionally the absurdity, of documentary "talking
heads" and the pretense of their detachment.[47] Using a "correct" ac-
cent (Received Pronunciation) and polite terms ("excreta"), Greenaway's erudite but stilted Expert also serves as a foil for the anonymous characters who speak far more easily about their relationships

Figure 11.6. "H is for Hi-Tec" still from *26 Bathrooms*. (Courtesy of Peter Greenaway)

to their bathrooms. However mundane, their comments convey the intimate texture, rhythm, sounds, and rituals of daily toilet use lost or passed over in professional discourse.

Greenaway's fascination with water has long been the subject of critical comment. Bathrooms, pools, and fountains play an important role in his films and in his curation of exhibitions such as "100 Objects to Represent the World" (1992), in which object number 49 is a bath. For Greenaway, according to critic Alan Woods, water stands for the various fluids that pass in and out as the body functions through the life cycle: giving birth, copulating, eating, and of course, excreting.[48] Memorably, in his *The Cook, the Thief, His Wife, and Her Lover* (1989), the first extramarital sexual encounter of Georgina (Helen Mirren) takes place in the restaurant bathroom. We might see *The Cook*'s bathroom as a site that combines both dirtiness (excrement) and naughtiness (illicit sex)—the ultimate in abjection—except that Georgina's affair marks the real beginning of her resistance to her gangster husband, Spica, and leads to her rebirth, if not her redemption. In this sense, the bathroom emerges as a sort of sacred site, as architect Marco Frascari defines it. One of the few architects to have written on bathrooms in contemporary design, Frascari invokes divinity in describing them as "numinous"— sacred places that permit transcendence, noting, "we walk in unclean (in Italian, rhetorically: *immondi*) and walk out clean (again, in Italian: *mondi*)."[49]

In *26 Bathrooms*, bathroom spaces do feel "numinous," but the transcendence on offer is not divine. For the most part, Greenaway keeps viewers grounded in the phenomenological. We see many nude bodies and are vicariously aware of the pleasing sensation of skin meeting water, skin meeting skin (e.g., a mother and her child), and skin meeting the ceramic or porcelain encasing of the bath; some of the film's characters are also shown enjoying the spectacle of their own bodies in the bathroom's mirrored surfaces. What the bathroom primarily appears to transcend is productive time. While it permits the activities necessary for one's participation in public regimes, such as work or school (e.g., "D is for Dental Hygiene"), it is also a space that stands outside them, a place of contemplation and relaxation, whose

Figures 11.7 and 11.8. "X is for an Expert on Bathrooms" still from *26 Bathrooms*. (Courtesy of Peter Greenaway)

dreamlike qualities are conveyed by the music or the sounds of water which waft in and out, carrying us along. Bathrooms thus emerge as routine *and* renewing, spaces of unmediated bodily pleasure, intimacy, and care.

In Greenaway's treatment, bathrooms also become hubs of social activity, a function that is reflected in their style and décor. In one scene, "U is for Underused Living Space," the character speaks of the need to have books and furnishings in bathrooms, and many of those featured in the film do. As well as being filled with armchairs, bookshelves, and artwork, they are lovingly made—Italian craftsmen are often and appreciatively mentioned—with salvaged bronze, glass, and porcelain features and hand-painted tiles. These decorated lived-in spaces bear little resemblance to the manly compact ones that Giedion celebrates or the ergonomic ones proposed by Kira. Insofar as they come from any tradition, it is that of the luxury English bathroom, where fittings are arranged "loosely" like furnishings.[50] Throughout the film, family members and sometimes friends gather in these commodious rooms to converse and to catch up on the "scandals," as one puts it, in a manner that evokes premodern traditions of public bathing (the public baths and communal toilets of ancient Rome come to mind), where cleansing and socializing activities intermingle, as described by Zena Kamash in chapter 3 of this volume.

Greenaway's aim, however, is not to convince us of the benefits of more public facilities: in "P is for Public Baths," an elderly woman's weekly visit to the baths is clearly viewed as a ritual from times past. Instead, Greenaway seems to want to remind us once again that, whatever Experts might say, bathrooms need not be treated only as contained, rational places; they are excessive by nature and sites of many potential pleasures and uses. In this regard, he is ultimately much more optimistic than Sigfried Giedion, who ends his celebration of the bath's formal qualities with the sobering conclusion that they actually express a crisis facing industrial society itself—the lack of a regenerative culture.[51] By contrast, Greenaway suggests that in the bathroom regenerative culture, plainly defined by Giedion as "time to live," remains within our grasp.[52]

The progressive transformation of the Expert over the course of the film seems intended to drive home this point. In contrast to the neutral living-room setting of the Expert's first two scenes, Greenaway later places him in an actual bathroom (fig. 11.8). Awkwardly wedged in this tiny space between a sink and a radiator, an Andrea Palladio poster at his back, the Expert's appearance of detachment becomes less credible, as do his claims for functionalism. As we hear the Expert in a voice-over defending mirrors as a "functional" design element (on the grounds that it is "very convenient" to see oneself "from all angles"), we watch a woman openly and in no way functionally relishing her nakedness in a mirror. With the next scene, any last vestige of objectivity is shed, as the Expert, now balanced on the bathtub rim, reveals that the bathroom in which he is positioned is his own. With this move, it is no longer possible to strictly segregate the two modes of talking about toilets—one public or official, the other personal and anecdotal—or to see them as operating in different registers. Rather, as the Expert himself becomes a user, the two modes begin to intertwine.

The Expert demonstrates his bathroom's various design features, including a porthole whose cover can be raised and lowered to modulate light and privacy levels. With something like glee, he tells an anecdote about having parties at his flat: when visitors close the porthole for privacy, the act of lowering the mechanism actually exposes them, alerting the company that they are inside. At this moment, the technology of the toilet begins to blend with the texture of its experience as the Expert "hooks" the user back up to bathroom equipment.[53] Here, as throughout the film, by deliberately drawing attention to the specific bodies that initiate use, Greenaway destabilizes the representational codes that govern text and image and resurrects what functionalist discourse generally tries to suppress: the individual, the erotic, the ironic, the everyday, the decorative, the dusky, the dirty, the fragrant, and the feminine. But one also senses that Greenaway quite simply relishes the act of "releasing" the Expert (a proxy here for Kira?) from the constraints of his talking-head role.

In its tone, the scene with the Expert also seems to presage another calculated and comic riposte to elitist discourse, John Berger's "Muck and Its Entanglements," a short article about cleaning his outhouse.

This piece moves between a detailed discussion of the mechanics of the operation, with careful descriptions of the sight and smells of various kinds of shit, and a metaphysical contemplation of purity and evil. The way the mechanical and metaphysical "entangle" is humorously invoked when Berger writes, "Kierkegaard (Shit!—half a shovelful has fallen off) knew what he was talking about when he defined diabolic discourse as prattle." It is unlikely, of course, that Berger was actually shoveling shit at the moment of writing, but this passage is intended precisely to show how it is impossible to completely detach even our most high-minded ideas and metaphors from everyday acts and the realm of the senses. In our entanglement with muck, Berger tells us, we are inevitably reminded "of our duality, of our soiled nature and of our will to glory."[54] Like Greenaway in *26 Bathrooms*, Berger's point is that muck always exceeds the ability of reason to discipline it.

Coda

Since 1976, when Kira published the revised edition of *The Bathroom*, and 1985, when Greenaway directed *26 Bathrooms*, much has changed, academically, socially, and environmentally. To the despair of Roger Kimball and his neoconservative confreres at the *New Criterion*, poststructuralist approaches have come to dominate academic activity at many universities. Serious analyses of public toilets are no longer so thin on the ground, and thanks to edited collections such as this one, academics from a spectrum of disciplines who have been working on the subject individually are beginning to get the sense of its belonging to a field. (Witness the success of the multidisciplinary "Outing the Water Closet" conference, organized by New York University and the American Institute of Architects, which gave birth to the present volume.) One gets the hopeful feeling that toilets are not as taboo as they once were.

But it is equally clear that toilet studies still need to address several charged representational issues, not least the question of how to talk about toilet usage within architectural discourse. What language do we deploy for such a messy subject? The question remains a critical one because, as we have seen, language has powerful effects. The cleansing of language and of visual representation conceals the interconnectedness of all our actions: focusing on the object (production) may allow us to forget habits of use (consumption) and, crucially, to ignore how they are bound in a cycle.

Yet disciplinary habits die hard. Even when we consciously resist the whitewashing of academic or of expert language, as Berger does, it can seem odd to find practical discussions of toilet matters alongside critical or metaphysical ones, the jostling between high and low often setting into play the same inner tension—between dirt and purity, transgression and control, private and public—which is analyzed within the piece. I suggest that this tension can be extremely productive for scholars, however, reminding us of the ways our own "body" of criticism is itself shaped by its objects of engagement (how we are ourselves "entangled") and demanding that we pay more attention to our writing and visual strategies when publicly dealing with dirty subjects.

Some scholars such as Berger or the anthropologist Sjaak van der Geest have dealt with this challenge by adopting the first person and a more intimate or memoiristic approach in order to acknowledge that, in writing about the toilet's history or mores, our own bodies, behaviors, and beliefs are also implicated.[55] Many of those writing about toilets also very deliberately reject euphemisms. Some do so in a spirit of Duchampian provocation.[56] Others, such as Rose George, deliberately choose plain words—*shit* over *excreta*, for instance—in order to resist the confusion or elisions that euphemisms create.[57] As George so persuasively shows, the refusal to deal openly with the realities of toilet use can have calculable and devastating impacts on local ecosystems, health, and living standards in developing countries. But she also provides abundant proof that we in the so-called civilized countries suffer from this blinkered approach as well.[58]

Nothing more convincingly makes George's point than the ongoing failure of local governments in the West to adequately and imaginatively address the issues of toilet provision, planning, and design, despite the fact that, when consulted, members of the public consistently identify decent and safe toilets as essential for well-being, dignity, and health.[59] It seems that users' preferences, requirements, and habits remain low on agendas that are driven largely by the fight against vandalism and shrinking budgets. But as Kira stressed, "a society can have anything it cares enough about and is willing to pay for."[60] With these words, he wisely reminds us that the battle to improve toilet provision will ultimately be settled in the field of representation: if the taboos surrounding serious discussions of toilet use can be overturned, the political and professional will to improve design or to think of it more integrally will follow. And if Kira and Greenaway have anything to teach, it is that architectural discourse and urban environments will be not only healthier but qualitatively richer when they come entangled with a user.

Rest Stop

Toilet Bloom @ Bryant Park

Foyer of Bryant Park restroom, New York, 2008.

THIS LANDMARKED BUILDING in Bryant Park near Times Square houses an exemplary public restroom. The long-dormant facility reopened in the 1990s after rehabilitation by the Bryant Park Restoration Corporation (BPRC)—a business improvement district created under the sponsorship, among others, of the Rockefeller Brothers Fund. Current BPRC director of operations Jerome Barth believes, "If they can do it at the Four Seasons, there's no reason we can't do it here." Indeed where else but luxury hotels and clubs are courteous male and female attendants present to keep the facility clean, well stocked, and freshened with elaborate flower bouquets and "aromatherapy"?

Toilets are equipped with sensors that trigger a revolving plastic sheath to guarantee a clean seat. Floors are tissue free, in part because attendants continually sweep up any refuse.

Exterior of Bryant Park restroom, New York, 2008.

12

On Not Making History

What NYU Did with the Toilet and What It Means for the World

Harvey Molotch

BUILDING A BETTER bathroom means approaching fundamental human disgusts and anxieties. Putting toilets into presence, whether in life or text, amplifies whatever distinctions and worries that are around. People do not want to know even that others have sat on the seat they occupy, much less visualize (or discuss) what those others have done or how to arrange for them to do it differently. That those others are in some way diverse or strange, by gender, class, or ethnicity, makes their shit even more worrisomely obnoxious. Hindus and Muslims won't use the facilities touched by the Other and stand ready to deprive them, when opportunity arises, of appropriate access. The rich avoid defilement from the poor, and heterosexual males do not want to have their pants down where homosexuality has been in play. Avoidance invites betrayal of the human rights of fellow beings made into dirt.

There are a lot of good reasons to question even sacrosanct practices, gender segregation most prominently. The biological differences once a basis for assigning women to specific roles and physical places have become obsolete and, in retrospect, ridiculous. This history should make us suspect the "necessity" of toilet segregation—this practice that "gives our public life a kind of segregative punctuation"[1] despite almost continuous contact between the sexes. Not just a matter of "attitude," reinforcement comes from diverse mechanisms such as building codes, bureaucratic organization, and news media reactions. Women join men in sustaining the pattern, both in their expressed desire to be with their own kind and the very act of selecting the "right" door from the forced-choice option in the hallway. In and out, in and out, with each swing reinforcing the binary. People who misplace themselves by choosing "wrong" become a walking scandal, a performance of incivility and threat. As in other realms, patterns of life and hardware take on a momentum of their own with an inertia of continuity regardless of how little sense they may make.

Even in realms less heavily freighted with apprehension than restrooms, innovation does not come easily. A whole lot of stars need to be aligned. The sociologist Bruno Latour has researched an infrastructural failure in his native France to determine what it takes. He examined the life (and death) of a proposed radical new train for Paris, called "Aramis." It involved automated small cars that would allow, in effect, the routing of train segments to specific destination choices of passengers. It was a sufficiently appealing idea to cause the French government to invest about $100 million in its planning and in a work-through of costs and benefits. Latour concluded, after extensive study of the actors and institutions involved, that there was not enough "love"—the necessary ingredient and the word he used in the subtitle of his book—to bring this new technosocial phenomenon into being.[2] It is "love," explains Latour, which determines whether or not something socially and physically complex can come into the world. The new train required a buy-in or, in Latour's vocabulary, "enrollment," from a diverse set of actors and a willingness to take risk for that which had not been in prior existence. Love operates as coordinating device to keep heterogeneous actors consistently on task.

So it was with the effort to create a new kind of public toilet at New York University, where I teach—a project that was the instigating effort for this book. There was not enough love. And in seeing what kept the love at bay, many of the lessons of the preceding chapters come to the fore. Understanding what went wrong gives us a route into what makes toilet change difficult, even under what appears to be propitious circumstances for transforming how things are done.

The new facilities were to be for the university's Department of Social and Cultural Analysis (SCA)—one of the two departments with which I am affiliated. We were allocated an entire floor (about sixteen thousand square feet) of a former industrial building, just down the block from the historic Cooper Union and adjacent to headquarters of the *Village Voice*. Everything was to be gutted and then remade according to the most forward-thinking design, ecological principles, and technological advance. Budget was more than adequate to provide for the approximately fifty offices, plus lounges, conference rooms, and other paraphernalia of an academic department that also sponsors a large number of public events in downtown New York. Eyes were wide at the prospect.

All six of SCA's constituent programs were born out of a post-'60s zeal for social change and diversity. Coming together in the new space were the once separate programs of Metropolitan Studies (my primary affiliation), Africana Studies, Asian/Pacific/American Studies, Gender and Sexuality, Latino Studies, and American Studies. The department's members proudly press the envelope against most all intellectual convention, and there are many gay and lesbian faculty, staff, and students; and some transgender individuals as well. Enthusiasms ran high for using physical apparatus to inscribe new visions of gender, sexuality, and access into the environment. We had the additional advantage of having as head of the campus's design and architecture department a woman of great imagination and strong sympathies in our direction. As a younger architect, she had spent part of a fellowship year abroad photographing public toilets in famous buildings such as the Louvre. She was "toilet," in the accolade we came to use for those who got it.

At this writing, the spaces have been built, and the men's and women's rooms are largely conventional, two separate rooms, both with sinks and stalls, one with urinals too. There was, in response to our initiatives, a decision to create a third space, a roomy single-occupancy toilet-sink combination accessed by its own entry. We also got the innovation of using the word "Toilet" (in large and bold lettering) to designate where the men's and women's rooms were to be found—no demure "restrooms" for us! But any more ambitious scheme failed to make the grade.

Administrative Reluctance

Administrators—including those at a large university, whether a public one such as the University of Massachusetts or a private one such as NYU—do not, as the stereotype of the gray bureaucrat would imply, daringly jump into something new. In creating new buildings and remodeling old ones, there is a set pattern for how to do things. Staying on pattern allows a lot of people to perform their duties across a range of specialties. Everyone "knows" what a building restroom should be like, that it will involve toilets and sinks, signs and separations, some spaces with urinals and some not. Access to plumbing of specific sorts at precise locations needs to be designed or captured in already existing facilities. To innovate means going back to the drawing boards, rethinking architectural opportunities and constraints, and checking continuously to make sure everyone is aware of the plan now being implemented. This is a hassle, one with financial implications and new potentials for error.

Even talking about toilets, besides the raw taboo against doing so, is a diversion from normal topics that do need to be taken up. Time spent on the toilet issue means time not spent on the basic and generic process of programming and designing the space. We spent hours figuring out how to maximize natural light, how to expose the building's old columns, how to balance private offices with shared spaces, and even how to provide chalkboards for self-expression in public areas. Working through details of restroom innovation was an extra, one that burdened an already crowded agenda. So whereas in less liberal environments any discussion of toilet innovation may be forbidden, even in a place like NYU it is hard to keep it on the agenda in a way that facilitates the kind of hard effort that accompanies reworking a deeply rooted set of design assumptions.

Toilets bring the further potential for institutional humiliation in the form of negative public reaction. One precedent for ridicule was the kind of publicity that accompanied another of my efforts to take the public restroom seriously. A journalist for the *New Yorker* magazine attended a session of my senior-level undergraduate seminar,

"The Urban Toilet," that I taught in the spring semester 2008. The *New Yorker* journalist spoke with me at some length and then published a short essay in the magazine's "Talk of the Town" section under the title "Powder Room 101." He wrote that the course's "remarkable syllabus reads almost like a parody of Allan Bloom's worst nightmare, bringing the jargon of gender and ethnic studies, city planning, and industrial design to bear on the most euphemized of subjects."[3] That syllabus bears a lot of resemblance to the chapters in this book; indeed, several of the authors are the same.

The *New Yorker* writer went on to report, "The fourteen students took turns describing their proposed field studies, which ranged from 'the idea of the cabdriver: when and how?' to a comparison of Starbucks bathroom use across neighborhoods, and a survey of the people's vision of the ideal commode."[4] Two of these very projects are among the works cited in the introduction to this volume—unusually useful among undergraduate term papers I have come across in forty years of teaching. And there were other good ones, too.

In class immediately after the *New Yorker* piece appeared, I began by asking the students what they thought of the coverage. A young woman summed up her surprise: "He didn't get it." In fact, I think this highly intelligent reporter understood everything. He had retreated to the default position on toilets as joke. Imagine what might occur if actual physical toilets had been installed reflecting such ideas. Less liberal outlets than the *New Yorker* would have a field day with the shenanigans of a university environment built to favor transgender people and other queers—something perhaps like the way conservative media at the University of Massachusetts reacted to the prospect of unisex toilets at that institution.

The administrators probably kept in mind that they are stewards of buildings that may well change in the departments they will house. Buildings do, as Stewart Brand famously wrote, change the kind of functions they serve and types of people who use them. They "learn," in his terminology, and that means what is good to do for an initial set of users may not be right for the next set.[5] Even under SCA, some of the visitors to the building will be people who were not aware of, let alone involved in, this inventive planning and whose sensibilities should be taken into account. Unlike in art installations or avant-garde behavioral breaches of social routine that the foolish or adventurous may carry out, public restrooms require physical investments that are expensive and hard to reverse. This adds to the quotient of needed boldness, including willingness to risk mockery, if change is to occur. At the same time, buildings are—virtually by definition—a history of prior "fixes" that were inventive at a prior moment. They display priorities and the elements of a new consensus that enabled them to come into being.

Building Codes and Higher Authorities

An organization such as a university is itself embedded in a hierarchy of authorities and officials with which to contend; they also weigh in. The New York City Department of Buildings has codes and standards for everything to do with construction, and that includes all matters of plumbing and restrooms. Any departure from the usual set-up of completely separate men's and women's rooms requires an exemption. And such exemptions probably involve others, such as numbers of stalls, of urinals (if any), of sinks and pressures, pipes and fittings. Exemptions do get made. For example, at the newly rebuilt Museum of Modern Art (MoMA), restrooms serving the posh restaurant, "The Modern," are unisex. Women and men (including those with disabilities) share a common bank of sinks and make use of single-occupant toilet stalls. Perhaps because of the gender sharing—and need to exclude smells and sounds coming from a differently sexed person—the enclosure walls are floor to ceiling (something common in Europe, where gender sharing is also more prevalent). So unisex of this sort, and ways to handle it, is a precedent coming from one of the city's most visible and elite institutions.

For any business or organization, negotiating exceptions takes time and effort. Construction occurs on a schedule, with design only an early part of a process with many steps. Plans must go to the city, be approved, and then come back to go out for bid. Time is money. In this case, since the university had purchased a building to remodel for SCA and other campus units and emptied it out, it was now—in effect—burning cash as the process moved forward. SCA had planned to be out of its prior leased space on a certain date; plus the faculty and staff were anxious to move into the new, superior digs. Gaining the exemption risked delay. Given the prospect of having to redesign after the fact, the university at one point instructed its SCA architects to draw up plans that assumed conventional design as well as a MoMA-like unisex plan—an added work effort.

The Department of Buildings (DOB) indeed got involved and in fact turned the university down. Toilets would be normal. An appeal was put forward but turned down again. Apparently the city's building commissioner expressed, according to our NYU administrative contact, "concerns about security and liabilities" created by going unisex. The city was fresh from a series of scandals involving falsification of building inspection reports that became linked, in the media, to crane accidents resulting in deaths. Indeed the building department has "a long history of graft and corruption," as a *New York Times* article headline reported.[6] It somehow followed, according to an email sent to us by the university official involved in the negotiations, that "it would be very difficult for anyone at the Department of Buildings to allow/permit any issues that could be interpreted as controversial and that are not currently covered in the NYC Building Codes, past or present." It would be, our source indicated, especially difficult to gain the needed permission "under the current environment at the DOB and in light of so many recent adverse events in the construction industry in New York City." The only way to make sense of the putative connection between the NYU restrooms and these matters to which they were somehow related is to think that mixing genders adds further matter out of place to the dirt of construction fatalities and corruption scandals.

Internal Politics of Sensitivity and Awareness

As also made evident in the chapters of this book, even progressive goals can be in mutual conflict (e.g., providing for the blind can conflict with providing for those in wheelchairs; providing for transgender people can be in conflict with providing for observant Muslims). As part of moving toward restroom innovation, the SCA department held a number of discussions among faculty and staff in workshop style to iron out any such divergences. And there was divergence. Some wanted a radical solution. That would mean retaining urinals for their space-saving and water-free ecological efficiencies in a single common, gender-neutral space. Men and presumably only men would use the urinals. Women and men would both use single-user stalls, built Euro-style with full-height partitions. Both men and women would, also as in the MoMA case, share the same sinks and mirrors.

But some women within SCA did not like the idea of being in restrooms with men, and some men had reservations about sharing with women. A number of women thought men were dirtier, based on experience at home or with single-user public facilities. Urinals do solve problems of men missing the bowl or not putting the seat back down (if they bother to put it up in the first place). But some women did not like the idea of being in the same room as men whose penises were out, regardless of the reason. They did not want to see the backs of men as they peed. Votes were held. The largest block of support was for a MOMA-like unisex layout, with no urinals but shared banks of sinks and individual toilet stalls. The radical solution garnered seven votes (mine among them). Five voters favored a completely conventional outcome.

There was a pattern to the internal split, one especially important in the SCA context. It was staff people who were most opposed to moving off convention. These workers are the individuals most dependent on the restroom, not faculty or students, who come and go. Staff remain in the confines of the department most of the day. At the time of planning for the new space, they were exclusively women. Most were women of color and, in terms of earnings and standard of living, had less capacities than the faculty, like me, who were filled with gender-ambitious plans. Those who keep the accounts, arrange the meetings, and prepare the curricula did not sign up for an experiment in gender relations. They signed up to do a job, to receive basic benefits, and to feel comfortable at work.

We arrive at still another intersection of race, class, and gender, all this in a department sharply attuned to any injustice on these charged fronts—and in this case, a conflict among heart-felt values. I tend to the view that, given the particular spirit of SCA, the staff would have adjusted and even become fond of the inventive set-ups that might have come about. The architects would have designed something to allay anxieties—as suggested in the graphic to follow reflecting our own schemed-up proposal for change. I am confident the male users would have been impressively solicitous of women in their restroom midst. Students and visitors would have taken note, maybe even, yes, have been stopped in their tracks. It would have been a teaching moment. Just as with some readers and at least some authors of this book as well, jokes and awkward tee-hees start the learning process, but more serious considerations follow. In a word: education.

Opposition was strong enough to eliminate working through creative architectural solutions. For example, urinals could have been in an alcove or behind a partition, perhaps one that exposed, to soothe security anxieties, men's feet and/or heads. Or there could have been a semitransparent scrim of some sort, or use of opaque glass, that could reveal anyone on the other side while otherwise obscuring body and organ shapes. Our prize-winning architects, a forward-looking design firm, Lewis.Tsurumaki.Lewis, were eager. They wanted to make restroom history. (Indeed, the finished project, although not the restrooms per se, won a "Best of Year" Merit Award from *Interior Design* magazine, which featured a multipage spread of SCA in its pages).[7]

As for those at SCA who would have remained displeased with the newly installed teaching aids, there would have been restroom alternatives as a matter of course. There were already conventional public restrooms on the floor above and the one below, occupied by other NYU units. (Indeed, in a practice that I believe to be quite common already, some people migrate to toilets at a distance, where their shoes are less recognizable.) In addition, all plans called for a single-user and lockable toilet-sink bathroom (open as to gender and disability and also for family use) directly off the main hallway of the department. This is especially of value to people who fear going into an open facility (men's or women's or both) when nobody seems to be around, especially when working late at night. Better to be able to open the door, see inside, and then lock the door behind you—an appropriate fallback alternative.

Despite the possible mitigations, reservations and complications cumulated. The department vote and the views expressed among staff and faculty did not support a consensus for radical change. And the city's building department would not even accept a variant of the MoMA format. I am of the conviction that the NYU administrators were not convinced either, and they did not have their heart in it— not enough love. I don't think they fought for the cause—a reticence this private university is not otherwise regarded as having when pursuing its ambitions, academic or otherwise. It was a lost opportunity to inscribe social change into architectural form and to use form to facilitate intellectual growth.

What Would a Better Bathroom Look Like?

If not for NYU now, then perhaps some other place at some new time. The NYU experience, combined with the reflections of the authors in this book, gives us the outlines for a utopian vision of men and women and people of all sorts sharing toilet space and shaping social life. We need a way to get parity for people who identify as males and females, as well as for those who do not clearly fit either category. And we want to provide options for people who need special facilities, in terms of space and equipment, without marking them off as segregated because of some kind of presumed ill fit. We need to accommodate people as they and their companions (including children and the aged) move through the world with their stuff—their baby carriages, crutches, wheelchairs, shopping wares, overcoats, insulin needles, and all the rest. And finally, we wish also to meet needs of those who, for cultural or utilitarian reasons, want alternative postures for defecation and ways to cleanse.

In trying to imagine what an ideal facility would look like, we do not have much help from visionary architects, especially the modernists, who often made such a fetish of showing it all—of designing buildings that expose their steel skeletons and utilities and in other ways being "honest" with the materials as well as the actual functions of their buildings. With the telling exception of Denise Scott Brown,[8] not coincidentally herself a modernist critic and an ardent investigator of the social ordinary, the important architectural figures stopped short of the toilet; boldness of vision failed them. So it is that, other than a few clubs for the transgressive young, restroom conventions remain intact. Restrooms are completely closed off from other parts of structures, without even a visual hint of what goes on within them. Little effort has gone into the kind of design, such as our SCA architects were willing to try, to invent new forms.

Laura Norén and I, amateurs though we are, have sketched out something we think moves the process forward. We've applied our suggestions schematically both to a high-volume restroom (fig. 12.1) like those at stadia and to a midscale facility like the one that might have been built in the office space at SCA (fig. 12.2). We follow in the footsteps of Clara Greed, who has, in one of her publications, provided an earlier blueprint.[9] Of course, Alexander Kira, whose good works have been described by Barbara Penner in chapter 11 in this volume, has led the way as well. In the version we suggest, there is a more or less single space with separation by *function*. That way people can sort themselves out by the equipment they need rather than what they putatively *are*—a solution that indeed aims for the kind of progressive and humane architectural outcome urged by Ruth Barcan in this volume.

As we have drawn up our scheme, we presume men will tend to use the toilets nearest the urinals, and women the toilets farther away. So there may well be some sorting by gender through this nudge of appliances but not through signage, walls, or closed doors. Boundaries are fuzzy, not strict; our restroom designs capitalize on the "pretense" that we know people so often use to enact privacy when physical separations are lacking. As the sex ratio of users changes, one gender can spill over into the facilities ordinarily overselected by the other, while, as the need arises, managing glance in an appropriate way. And as social attitudes shift over time, the usage choreography can alter with it, with no need for physical reconstruction. This is flexible design.

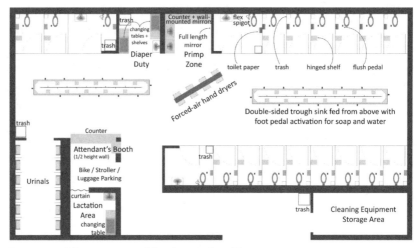

Figure 12.1. Schematic proposal for a large-scale public restroom.

The two regions meet in a quasi-lobby area, where mirrors and counters exist for grooming, in part to separate sinks from hair (we were told by an architect informant that long hair clogs sinks, so sinks and mirrors should not be colocated). The grooming area is also put at a distance from toilets to mitigate offending sounds and odors. Stalls would contain controls to set off sounds to mask human noises, should the human involved want this to happen (such is now an option in some Japanese toilets). One could also create white noise throughout the facility using music or newscasts or, perhaps to directly embrace the comedic, sounds of flatulence, pissing, and bowel elimination to obscure origins of the real thing. Of course, the "music" could be changed for different moods, times of day, and types of users.

Foot pedals would replace hand-operated flush controls in stalls and operate water flow and soap supply at the sinks. Pedals are already sporadically in use both in the United States and in other parts of the world. They conform to typical usage—many people kick to flush already—and they are hygienic without restricting user behavior, as electronic sensors do. Several stalls would provide for wheelchairs and also offer flexible spigots for cleansing in lieu of toilet-paper dependence. Although we did not detail them in this rendering, there should

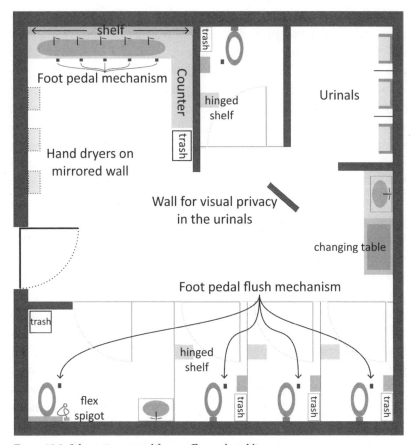

Figure 12.2. Schematic proposal for an office-scale public restroom.

also be stalls with squat-type toilets for those who prefer them and to encourage others to make the try. Further, we suggest one-way peep holes (mirrored on one side?) to enable those within the stall to see who is outside, both as a security reassurance and also to inform that someone may be waiting. Also at the bank of sinks should be at least one stepping stool so that children or very short people can step up and reach them. And there should always be tampons and menstrual pads available (maybe bandages as well); perhaps we have now reached a point of sufficient wealth and civility to provide them as a free good.

While we're at it, may we also suggest toilets that show what goes on—making excrement visible so that worms, undigested items, or other problem signs (especially important in the case of children's bowel movements) can be readily detected. Conventions of excrement display are culturally specific, as the Slovenian intellectual Slavoj Žižek has poignantly indicated.[10] At least before the most recent moments of hyperglobalization, difference persisted even within western Europe. French toilets tended to have a hole at the rear of the bowl that causes solid waste to disappear at the moment of release. German toilets have had a shelflike surface that displays excrement until flushing. And Anglo-American toilets held the stuff under water, a halfway solution. Especially with adequate ventilation, the German mechanism seems most wholesome, and our utopian vision assumes it.

In the high-volume restroom we have drawn up, there is an attendant to deal with people as they enter, to keep watch over items left in an anteroom area at the entrance, and, of course, to come to the aid of a person in distress. Our attendant provides real help—not like the perfunctory "service" of attendants in some posh places who go through the motions of turning on taps and handing out towels, probably causing some to skip the washup completely rather than face the social awkwardness or to avoid giving a tip. Our attendants practice security primarily through their helping. They might provide, for example, needle exchange to supplant dispensers in the stalls, perhaps even with timely advice about drug impurities in circulation and available counseling services for those so inclined (not so far-fetched given current practices in the Netherlands in coffee houses and marijuana shops). Integrated into the attendant's suite is a comfortable lactation facility.

Besides this approach of building fresh and new, there are simple and cheap solutions to deal with bathrooms already in place. One is for businesses and institutions to just take down the gender-specific signs and let people randomly choose their spots. Another is to remove the entry doors of all the restrooms, making them instantly unisex. There is also the possibility of bottom-up action—the toilet equivalents of a Rosa Parks or a Gandhi. Some women, when they find the long line at their restroom, head into the other one. I've heard it said that such women "take over" the men's rooms, a phrase I also used to hear applied to African Americans who had the temerity to move into a white neighborhood or "invade" a white drinking establishment. As still another route to insurrection, users can themselves tear down, paint over, or switch the signs from time to time. The guerrilla blurring might help enact changes in usage and custom or at least raise the topic.

The Ongoing Saga

To deal with waste intelligently, it has to become—to repeat the story still again—approachable. But so much of the civilizing process, designed to enact distance from the animal nature of what we are, stands in the way. In the Middle Ages, among the most widely read publications of the time were instructions for how to do the separation properly. So to deal with your snot, you need to keep fingers out of the nose (and ears) at mealtime, and one must observe the appropriate place to go for defecation and urination.[11] One should not, as given by a sixteenth-century German source, "relieve oneself without shame or reserve in front of ladies."[12] In what Elias called the historic "shifting frontier of embarrassment," we have moved to an era when such explicitness, even as proscription, is itself taboo. Even when—maybe especially when—toilets and plumbing became central in the national dialog, in the nineteenth century (see Kogan, chapter 7, in this volume), the etiquette books "had little, in fact almost nothing, to say on the subject of plumbing etiquette."[13] Hence, silence of elimination has become enshrined as a great historic accomplishment—a black box not to be opened.

To reverse course—to reopen civilization—it will take quite a lot. To process urine and shit intelligently, in both ecological and social terms, the "excluded monster"—as Michael Thompson calls waste more generally[14]—will need to be *tended*. In a way, the need runs parallel to the "slow food" movement, which also strives to increase awareness of one's own natural body and its relation to the world. Just as there now is a collective effort in the rich countries to take food seriously as nutrient and savor the time and effort it takes to properly prepare and consume it, we need to patiently attend to the other end of the process. Rather than feverishly rushing waste thoughtlessly away, it should be made fit for democratic purpose and ecological benefit. Slow Food, Slow Shit. A shift, for example, to dry toilet systems (that do not use water but turn feces into benign residue) requires periodic aeration of the solid mass, such as by rotating a drum or working an "aerator rake" through the mass. Facing up to waste can move feces and urine toward soil enrichment or energy source—and also toward solutions of everyday human problems of access and humiliation. At the level of national technology investment, appropriate and large-scale research and development at our great universities and elsewhere could make constructive sanitation cheaper and easier.

Challenging any effort at reform (slow or fast), as the chapters in this book repeat, is the way larger features of society impinge. Fear of others, by class or race or gender, interferes. Those who have options do not want to share with those who do not, either directly by giving them adjacent space or indirectly by expending resources to provide for needs not directly their own. And it may well be, as it has been in the past, that some people—either through attitude or practical circumstance, may well betray the niceties of life when left alone and unguarded. They may put aside concern for whatever and whoever comes next and vandalize or threaten in more direct ways. But again, phobia among the privileged is a longstanding tradition; the threat of the poor and the Other are vastly exaggerated and misplaced, something that those who study fear of crime and anxieties over risk repeatedly document.[15]

Part of conventional gender relations is women's fear of men—or the fear that men interpret them as having or, indeed, insist that they have. When combined with racial hate, stories of white women offended by black men fired up lynch mobs. Fear of men perpetuates odd and damaging arrangements. Men are not, as a rule, violent; increasing numbers of people in any given environment helps maintain social order—the famous principle of "eyes on the street" handed down from Jane Jacobs. In the case of SCA, to return to the learning experience at hand, the presence of many gay and lesbian people further alters the calculus. At SCA, sexual frisson (or threat) is as plausible in sex-segregated as in sex-integrated spaces. With the increasing acknowledgment of homosexuality in society, this is true elsewhere as well. As Mary Anne Case indicates in chapter 10 in this volume, hanging a sign on the door does not keep miscreants out. To whatever degree gender separation implies a false sense of security (and lessens it by lowering the number of people around), it blocks the unisex configuration that might alleviate not only the criminality problem but the other difficulties as well. Ironically, it might be especially useful to avoid gender segregation when men dominate in the environment because that is the most likely condition when a woman would find herself the only person in the women's room. Maybe imposition of gender segregation, aimed at enhancing security, undermines it—a recurrent issue raised by many security-oriented policies.[16]

The vexing dilemma for those trying to confront "antisocial" behavior at the toilet is that the privacy considered necessary for the benign person yields problems in trying to deal with those who create trouble. Surveillance clashes with privacy. Authorities, we might say, whatever their panopticon capacities, have trouble seeing through stalls. The little Harvard experiment of removing stall doors in the men's room (see "Rest Stop: Erotics at Harvard," in this volume) did not work. The doors were replaced before very long. The users were too collectively important to deprive of a place to go and perhaps also insufficiently homophobic to want to deprive themselves in order to keep men from having sex with each other. A more time-tested strategy is to install human inspectors, police who entice gays into displays of sexual intent. The hapless Senator Larry Craig was so entrapped at the Minneapolis airport, where vice-squad dicks evidently served on regular detail. Their peculiar presence is indicative of the zeal, even when there is physical and visual separation, to inspect and know. Such mixed commitments are true not just for the toilet but—to varying degree—wherever people are present but cannot be easily scrutinized and controlled.

The solution, an alternative for government as well as for individuals, is to accept the dictum that "shit happens," in both the literal and figurative sense. The dilemmas are best resolved by adapting to the fact that things can go wrong. Live with it. Businesses put up with some theft rather than scrutinizing every worker and customer. Some companies are at ease with a number of customers exploiting special advantage, even if it seems, at first glance, to weaken the bottom line. So Apple runs its computer stores with an array of free services such as classes, coaching, and simple advice (as did Singer sewing stores of an earlier time). The company presumes payoff in the long run—even if some of these services impose unusual time demands. And some people will never buy, which is also part of the deal. Prestige department stores take back merchandise that may have been used, even if it means having to destroy the goods as unsellable. Again, the system works best in a benign environment, one that makes people less desperate to game the arrangement, and where mutual good faith can be relied on. More democratic and egalitarian societies thus have the best chance of providing public goods with minimal liabilities, something that appears to be a general phenomenon.[17] But even when the costs are high, they should be paid, not only for benefit of those low on the totem pole but also for ease of the population at large.

Otherwise, the alternative is to close things down. To gain control over the few, authorities must deprive the many or engage in bizarre systems of inspection and control that betray, in their way, what many people would regard as true public decency. The public toilet displays an ultimate contradiction of capitalism—or of any regime made anxious by the exercise of autonomous capacity. We see how fear of others can put in jeopardy the capacity for achieving collective benefit—in this case, places to go for each according to her needs. The danger is all around us.

Notes

Notes to Chapter 1

1. My use of the term "bare life" differs from others, for example, Agamben, who uses the term to indicate a Foucault-like penetration of authority to the level of individual personality and biological being. See Giorgio Agamben, *Homo Sacer: Sovereign Power and Bare Life,* trans. Daniel Heller-Roazen (Stanford, CA: Stanford University Press, 1998).

2. This point derives from observations by Tamar Remz, whose NYU senior project involved a comparative census of an affluent New York City neighborhood with a poor one. Tamar Remz, "Counting the Spots," senior project, Metropolitan Studies, Department of Social and Cultural Analysis, NYU, May 2008.

3. American Restroom Association, "Fighting for Your Right to Use a Restroom," http://americanrestroom.org/pr/fightback.htm (accessed April 23, 2009).

4. Haegi Kwon, "Public Toilets in New York City: A Plan Flushed with Success?" M.S. thesis, Department of Urban Planning, Columbia University, 2005.

5. Kouteya Sinha, "Every Day 1.1bn People Poo without a Loo," *Times of India,* March 17, 2010.

6. Reports such as this are contained in the documentary film *Q2P,* by Paromita Vohra.

7. Rose George, *The Big Necessity: The Unmentionable World of Human Waste and Why It Matters* (New York: Metropolitan Books, 2008), 179.

8. Sulabh International Social Service Organisation, "Aims & Objectives," http://www.sulabhinternational.org/ngo/aims_objective.php (accessed August 5, 2009).

9. For a fuller discussion, see Barbara Penner, "Female Toilets: (Re)Designing the Unmentionable," in *Ladies and Gents,* ed. Olga Gershenson and Barbara Penner (Philadelphia: Temple University Press, 2009).

10. Harvey Molotch, "The Rest Room and Equal Opportunity," *Sociological Forum* 3, no. 1 (1988): 128–32.

11. Nicole Suarez, "Smoking in the Girls' Room: Toilets and Female Gender Formation," senior project, Metropolitan Studies, Department of Social and Cultural Analysis, NYU, May 2008.

12. For a senior thesis project at NYU, Kerri Berson took to campus buildings in a wheelchair to simulate the difficulties facing those with disability. Kerri Berson, "A Half Ass Job?" senior project, Metropolitan Studies, Department of Social and Cultural Analysis, NYU, May 2008.

13. Howard Becker, *Outsiders: Studies in the Sociology of Deviance* (New York: Free Press, 1973), 147.

14. Spencer E. Cahill, William Distler, Cynthia Lachowetz, Andrea Meaney, Robyn Tarallow, and Teena Willard, "Meanwhile Backstage: Public Bathrooms and the Interaction Order," *Journal of Contemporary Ethnography* 14 (1985): 33–58.

15. Santosh M. Avvannavar and Monto Mani, "A Conceptual Model of People's Approach to Sanitation," *Science of the Total Environment* 390 (2008): 1–12 (published by Elsevier, online at www.sciencedirect.com).

16. See, for example, Tim Ingold, *The Perception of the Environment: Essays in Livelihood, Dwelling, and Skill* (London: Routledge, 2000); Nigel Thrift, *Non-Representational Theory* (London: Routledge, 2008).

17. Galen Cranz, *The Chair: Rethinking Culture, Body, and Design* (New York: Norton, 1998).

18. George, *Big Necessity,* 130.

19. Laud Humphreys, *Tearoom Trade* (Chicago: Aldine), 1975.

20. George B. Davis and Frederick Dye, *A Complete and Practical Treatise on Plumbing and Sanitation,* cited in Andrew Brown-May and Peg Fraser, "Gender, Respectability and Public Convenience in Melbourne, Australia, 1859–1902," in *Ladies and Gents,* ed. Olga Gershenson and Barbara Penner (Philadelphia: Temple University Press, 2009).

21. National Association for Continence, "What Is Incontinence?" http://www.nafc.org/bladder-bowel-health/; see also George, *Big Necessity,* 142.

22. I thank Guido Martinotti, University of Milan, Bicocca, and Eva Cantarella, University of Milan, Statale, for bringing this inscription to my attention. The inscription was uncovered by archaeologist Luciana Jacobelli at Stabia in the municipality of Castellammare, at a site dating from the Late Roman Republic.

23. "The Changing Face of Taxi and Limousine Drivers," Schaller Consulting Archive, http://www.schallerconsult.com/taxi/taxidriversummary.htm (accessed August 6, 2009).

Notes to Chapter 2

This chapter arose from an innovative conference organized by Harvey Molotch, of New York University, and the Center for Architecture, "Outing the Water Closet, Sex, Gender, and the Public Toilet," which aimed to bring together academics in sex, gender, and space with architects, designers, and planners. I would like to thank the conference organizers for their invitation to attend, and to acknowledge the role that the speakers, respondents, and audience had in helping me to think more fully about public toilets as designed realities.

1. Deborah Fausch, "The Knowledge of the Body and the Presence of History—Towards a Feminist Architecture," in *Architecture and Feminism,* ed. Debra Coleman, Elizabeth Danze, and Carol Henderson (Princeton, NJ: Princeton Architectural Press, 1996), 39.

2. Mary Douglas, *Purity and Danger: An Analysis of the Concepts of Pollution and Taboo* (1966; repr. London: Routledge and Kegan Paul, 1984), 115.

3. Ibid.

4. Elizabeth Shove, *Comfort, Cleanliness and Convenience: The Social Organization of Normality* (Oxford, UK: Berg, 2003), 148.

5. I would like to thank one of the respondents to this panel, Matthew Sapolin, Executive Director of the New York City Mayor's Office for People with Disabilities, for highlighting these functions of public toilets.

6. When I use the term *public* toilets, I mean a variety of toilet types—from shared spaces in public places such as railway stations to toilets in commercial spaces such as shopping malls. The British Toilet Association more accurately terms these "away from home" toilets. Quoted in C[lara] Greed, "The Role of the Public Toilet: Pathogen Transmitter or Health Facilitator?" *Building Services Engineering Research and Technology* 27, no. 2 (2006): 1. Much

of what I have to say about the public performance of intimacy pertains to *shared* away-from-home facilities (rather than to single-user spaces such as toilets in a small café or restaurant), though many of the reflections on dirt are also relevant to these types of spaces. It is also important to note here that my comments are based on my experiences in Australia. Each country—indeed, as my experiences at the New York conference taught me, even each city—differs in the nature and extent of public toilet provision, with unisex facilities being much more common in some countries than in others, and with some countries having a far greater provision of public facilities than others do. This is true even across relatively congruent cultures (the United States and Australia, for example), so it is obvious that there are enormous differences on a global scale. My chapter, while it may contain some general theoretical propositions or observations that can be considered in a variety of different regional or national contexts, does not pretend to address anything beyond my own limited cultural context.

7. George Lakoff and Mark Johnson, *Metaphors We Live By* (Chicago: University of Chicago Press, 1980).

8. Norbert Elias, *The Civilizing Process: The History of Manners and State Formation and Civilization*, trans. Edmund Jephcott (Oxford, UK: Blackwell, 1994), 95.

9. Ibid.

10. Ibid., 94.

11. Alan Hyde, "Offensive Bodies," in *The Smell Culture Reader*, ed. Jim Drobnick (Oxford, UK: Berg, 2006), 54.

12. Ibid., 56.

13. Elias, *Civilizing Process*, 493, 497.

14. Susan Bordo, "Reading the Male Body," *Michigan Quarterly Review* 32, no. 4 (1993): 696–735.

15. The hygiene hypothesis is one name for the increasingly popular theory that the increasingly "clean" lifestyle of Westerners robs the immune system of its chance to be exercised and is implicated in the rise of allergies and so-called superbugs. See, for example, University of Michigan Health System, "The Hygiene Hypothesis: Are Cleanlier Lifestyles Causing More Allergies for Kids?" *Science Daily*, September 9, 2007, http://www.sciencedaily.com/releases/2007/09/070905174501.htm. If this hypothesis proves true, then it invites another shift in our definition of *cleanliness*.

16. Joel Sanders, introduction to *Stud: Architectures of Masculinity* (New York: Princeton Architectural Press, 1996), 17.

17. Hyde, "Offensive Bodies," 53.

18. Barbara Penner, "Female Urinals: Taking a Stand," *Room 5*, no. 2 (2001): 24–37; see also Penner, chapter 11, in this volume.

19. Harvey Molotch, "The Rest Room and Equal Opportunity," *Sociological Forum* 3, no. 1 (1988): 128–29.

20. These interviews were carried out as part of a broader study on nudity. See Ruth Barcan, "Privates in Public: The Space of the Urinal," in *Imagining Australian Space: Cultural Studies and Spatial Inquiry* (Nedlands: University of Western Australia Press, for the Centre for the Study of Australian Literature, 1999), 75–92; and Ruth Barcan, "Dirty Spaces: Communication and Contamination in Men's Public Toilets," *Journal of International Women's Studies* 6, no. 2 (June 2005), http://www.bridgew.edu/SoAS/jiws/Jun05/index.htm.

21. T. S. Eliot, "The Love Song of J. Alfred Prufrock," in *The Complete Poems and Plays: 1909–1950* (New York: Harcourt, Brace, 1952), 4.

22. It is interesting to speculate about which beauty procedures *cannot* be performed in this space. Mouth gargling? Waxing upper lip? Tweezing chin hairs? Flossing? All these are possible in practical terms, requiring little time or equipment, but are culturally less likely, for reasons that might have something to do with embarrassment thresholds.

23. Beverley Skeggs, "The Toilet Paper: Femininity, Class and Mis-Recognition," *Women's Studies International Forum* 24, nos. 3–4 (2001): 302.

24. Ibid.

25. Jacques Lacan, "The Agency of the Letter in the Unconscious or Reason Since Freud," in *Écrits: A Selection,* trans. Alan Sheridan (New York: Norton, 1977), 151.

26. Park is the Chair of the New York Association for Gender Rights Advocate (NYA-GRA). Her remarks were made at the "Outing the Water Closet" conference at the Center for Architecture.

27. Gay Hawkins, "Down the Drain: Shit and the Politics of Disturbance," *UTS Review* 7, no. 2 (November 2001): 32–42.

28. Jo-Anne Bichard, Julienne Hanson, and Clara Greed, "Please Wash Your Hands," *Senses and Society* 2, no. 3 (2007): 385–90.

29. McKinney is one of the main people in charge of public toilets in New York's parks. He made this comment as part of his response to a panel of speakers at the "Outing the Water Closet" conference.

30. C. P. Gerba, C. Wallis, and J. L. Melnick, "Microbial Hazards of Household Toilets: Droplet Production and the Fate of Residual Organisms," *Applied Microbiology* 30, no. 2 (1975): 229.

31. Gerba is the author of over four hundred articles on environmental contamination. A sampling of his work confirms the anthropological insight that cultural factors play a major role in our perception of what is clean and unclean. Customer-shared pens, for example, which seem quite "clean," culturally speaking, have only marginally fewer protein and biochemical markers on them than public bathroom surfaces, which in turn have substantially fewer than playground equipment or bus rails and armrests. Kelly Reynolds et al., "Occurrence of Bacteria and Biochemical Markers on Public Surfaces," *International Journal of Environmental Health Research* 15, no. 3 (2005): 230.

32. Charles P. Gerba, "Application of Quantitative Risk Assessment for Formulating Hygiene Policy in the Domestic Setting," *Journal of Infection* 43 (2001): 95.

33. Charles P. Gerba, "The Significance of Fomites in Disease Transmission and the Role of HACCP," PowerPoint presentation, Departments of Soil, Water and Environmental Science and Epidemiology and Biostatistics, University of Arizona.

34. Sheri L. Maxwell and Charles P. Gerba, "Men vs. Women: Office Study," report, Department of Soil, Water and Environmental Science, University of Arizona. December 14, 2006.

35. The public restroom tap is a relatively low source of germs, having under half the number of coliform bacteria found on the floor in front of the toilet—a site that has over twice the coliform bacteria found inside the urinal itself. Interestingly, cultural and microbiological ideas of contamination seem to match up in the case of the sanitary-napkin disposal bin, which rates high in both symbolic and literal terms. For menstruation as a polluting sign of sexual difference, see Julia Kristeva, *Powers of Horror: An Essay on Abjection,* trans. Leon S. Roudiez (New York: Columbia University Press, 1982).

36. I am using this example from Nicholson Baker's delightful novella *The Mezzanine,* in which he describes the protagonist Howie's awkwardness about the enforced sociality of the

office toilets: "I used the stalls as little as possible, never really at ease reading the sports section left there by an earlier occupant, not happy about the prewarmed seat." Nicholson Baker, *The Mezzanine: A Novel* (New York: Vintage Books, 1986), 83. (See Barcan, "Dirty Spaces," 12.)

37. Gerba, "Significance of Fomites." To be fair, Gerba's work also makes it clear that one of the reasons that home is where the germs are is because daycare is where the germs are. Children's playgrounds are also identified as places with high rates of bacterial and biochemical markers (Reynolds et al., "Occurrence of Bacteria").

38. Kelly R. Bright et al., "Heterotrophic Bacterial Levels on Common Workplaces Surfaces," unpublished paper; Maxwell and Gerba, "Men vs. Women."

39. Gerba, "Significance of Fomites."

40. This may be changing with the advent of environmental fragrancing as a marketing device. See Peter Damian and Kate Damian, "Environmental Fragrancing," in *The Smell Culture Reader*, ed. Jim Drobnick (Oxford, UK: Berg, 2006), 148–60.

41. Constance Classen, David Howes, and Anthony Synnott, *Aroma: The Cultural History of Smell* (London: Routledge, 1994), 5.

42. David Howes, "Hyperesthesia, or, The Sensual Logic of Late Capitalism," in *Empire of the Senses: The Sensual Culture Reader*, ed. David Howes (Oxford, UK: Berg, 2005), 290.

43. Classen, Howes, and Synnott, *Aroma*, 5.

44. Ibid., 4–5.

45. Eric Michaels, *Unbecoming: An AIDS Diary* (Rose Bay, Australia: EmPress, 1990).

46. Ibid., 42.

47. Michael Bull and Les Back, "Introduction: Into Sound," in *The Auditory Culture Reader*, ed. Michael Bull and Les Back (Oxford, UK: Berg, 2003), 8.

48. Kristeva, *Powers of Horror*.

49. Molotch, "Rest Room and Equal Opportunity," 130.

50. Bull and Back, "Introduction," 5.

51. Bryan Reynolds, "Sexuality and *Appendx*: Doorless Toilet Stalls and the Constipation of Desire," *Appendx* 1 (1997), http://projects.gsd.harvard.edu/appendx/dev/issue1/Reynolds.

52. Lee Edelman, "Men's Room," in *Stud: Architectures of Masculinity*, ed. Joel Sanders (New York: Princeton Architectural Press, 1996), 150.

53. Ibid., 153.

54. Hawkins, "Down the Drain," 36 (emphasis in original).

55. Hyde, "Offensive Bodies," 56.

56. Ibid.

57. Ibid.

58. Ibid.

59. Ibid., 57.

Notes to Chapter 3

Particular thanks go to Gemma Jansen for sharing her thoughts on the art historical and epigraphic evidence found in Roman latrines, as well as to the wider group of latrine specialists who attended the Ancient Latrine Workshop in Royal Dutch Institute in Rome in June 2007. Thanks also to Jack Tannous for finding and translating the Syriac references. The audiences at the London School for Tropical Diseases and the Roman Discussion Forum in Oxford made useful comments on earlier versions of this work.

1. Richard Stillwell, *Antioch-on-the-Orontes III* (Princeton, NJ: Published for the Committee by the Department of Art and Archaeology, 1941), 21.

2. Mary Phillips et al., "Disgust: The Forgotten Emotion of Psychiatry," *British Journal of Psychiatry* 172 (1998): 373–75.

3. Valerie Curtis and Adam Biran, "Dirt, Disgust and Disease: Is Hygiene in Our Genes?" *Perspectives in Biology and Medicine* 44, no. 1 (2001): 17–31.

4. Junichiro Tanizaki, *In Praise of Shadows,* trans Thomas J. Harper (New Haven, CT: Leete's Island Books, 1977).

5. Michael L Satlow, "Jewish Constructions of Nakedness in Late Antiquity," *Journal of Biblical Literature* 116, no. 3 (1997): 429–54; Stefanie Hoss, *Baths and Bathing: The Culture of Bathing and the Baths and Thermae in Palestine from the Hasmoneans to the Moslem Conquest; with an Appendix on Jewish Ritual Baths (Miqva'ot)* (Oxford, UK: British Archaeological Reports, 2005), 12.

6. Seneca, *Naturales Quaestiones,* 1.16.4; Carlin Barton, "Being in the Eyes: Shame and Sight in Ancient Rome," in *The Roman Gaze: Vision, Power and the Body,* ed. David Fredrick (Baltimore: John Hopkins University Press, 2002), 217.

7. Jim Drobnick, ed., *The Smell Culture Reader* (Oxford, UK: Berg, 2006), 1–12.

8. Ibid., 15; Gale Largey and Rod Watson, "The Sociology of Odors," in *The Smell Culture Reader,* ed. Jim Drobnick, 29–40 (Oxford, UK: Berg, 2006); Donald Tuzin, "Base Notes: Odor, Breath and Moral Contagion in Ilahita," in *The Smell Culture Reader,* ed. Jim Drobnick, 59–67 (Oxford, UK: Berg, 2006).

9. Helen Chapman Davies, *The Archaeology of Water* (Stroud, UK: Tempus, 2008), 84.

10. Gemma Jansen, "Interpreting Images and Epigraphical Testimony," in *Roman Toilets: Their Archaeology and Cultural History,* ed. Gemma Jansen, Ann Koloski-Ostrow, and Eric Moorman (Leiden: BABESCH, forthcoming).

11. Katherine Dunbabin, "*Baiarum grata voluptas*: Pleasures and Dangers of the Baths," *Papers of the British School at Rome* 57 (1998): 1–46; Zena Kamash, "Water Supply and Management in the Near East 63 BC–AD 636," D.Phil. thesis, Magdalen College and School of Archaeology, University of Oxford, 2006; Zena Kamash, "What Lies Beneath? Perceptions of the Ontological Paradox of Water," *World Archaeology* 2 (2008): 224–37.

12. Kamash, "What Lies Beneath?"

13. Pascal Boyer and Pierre Lienard, "Why Ritualized Behaviour? Precaution Systems and Action Parsing in Developmental, Pathological and Cultural Rituals," *Behavioural and Brain Sciences* 29 (2006): 1–56; Siri Dulaney and Alan P. Fiske, "Cultural Rituals and Obsessive-Compulsive Disorder: Is There a Common Psychological Mechanism?" *Ethos* 22, no. 3 (1994): 243–83; Alan P. Fiske and Nick Haslam, "Is Obsessive-Compulsive Disorder a Pathology of the Human Disposition to Perform Socially Meaningful Rituals? Evidence of Similar Content," *Journal of Nervous and Mental Disease* 185, no. 4 (1997): 211–22.

14. Andrew Wilson, "Incurring the Wrath of Mars: Sanitation and Hygiene in Roman North Africa," in *Cura Aquarum in Sicilia: Proceedings of the Tenth International Congress on the History of Water Management and Hydraulic Engineering in the Mediterranean Region,* ed. Gemma Jansen (Leiden: BABESCH, 2000).

15. Mary Douglas, *Purity and Danger: An Analysis of the Concepts of Pollution and Taboo* (London and New York: Routledge and Kegan Paul, 1966), 165; William James, *The Varieties of Religious Experience* (Cambridge, MA: Harvard University Press, 1901–2), 129.

16. Adam Goldwater, forthcoming chapter on northwestern provinces, in *Roman Toilets:*

Their Archaeology and Cultural History, ed. Gemma Jansen, Ann Koloski-Ostrow, and Eric Moorman (Leiden: BABESCH, forthcoming).

17. For a full discussion of these latrines, see Kamash, "Water Supply and Management," 179–83.

18. For example, Hoss, *Baths and Bathing.*

19. Gideon Foerster and Yoram Tsafrir, "Nysa-Scythopolis—A New Inscription and the Titles of the City on Its Coins," *Israel Numismatic Journal* 9 (1986–87): 53; Fergus Millar, *The Roman Near East 31 BC–AD 337* (Cambridge, MA: Harvard University Press, 1993), 378.

20. Ehud Netzer, *Hasmonean and Herodian Palaces at Jericho: Final Reports of the 1973–1987 Excavations,* vol. 1 (Jerusalem: Hebrew University of Jerusalem, 2001), 211–12.

21. Bar Hebraus, *Chronicon Ecclesiasticum,* col. 263.

22. François Nau, ed. and trans., "Notice historique sur le monastère de Qartamain," *Actes du XIV congres international des orientalistes* 2 (1907): 1–75.

23. Kamash, "What Lies Beneath?" 224–37.

24. Veronica Strang, *The Meaning of Water* (Oxford, UK: Berg, 2004), 77.

25. Carlin Barton, "Being in the Eyes: Shame and Sight in Ancient Rome," in *The Roman Gaze: Vision, Power and the Body,* ed. David Fredrick (Baltimore: John Hopkins University Press, 2002), 217.

26. Kamash, "Water Supply and Management," 224–29.

Notes to Chapter 4

I would like to thank Laura Norén and Harvey Molotch for their invaluable comments on this chapter. Special thanks to my father, Dan Braverman, a gastroenterologist in his heart and soul, whose enthusiasm and humor, especially on bathroom matters, have provided a strong foundation for this project. Research for this study was supported by the Baldy Center for Law and Policy at the University at Buffalo, SUNY.

1. Michel Foucault, *Discipline and Punish: The Birth of the Prison,* 1st American ed. (New York: Pantheon Books, 1977), 195.

2. Interview with Supervising Public Health Sanitarian, Erie County Department of Health, April 16, 2008.

3. In *Katz v. U.S.,* 389 U.S. 347 (1967) (Harlan, J., concurring), the U.S. Supreme Court established the linguistic formula of "reasonable expectation of privacy" regarding eavesdropping on a public telephone booth. Legal scholar David Sklansky has recently suggested that the subtext of *Katz*—and, in fact, the main reason behind this decision—was the protection of gay men from police surveillance in public restroom stalls. David Alan Sklansky, "One Train May Hide Another: *Katz,* Stonewall, and the Secret Subtext of Criminal Procedure," *UC Davis Law Review* 41 (2008): 875.

4. Interview with previous director of Genesee County's Health Department, May 7, 2008.

5. Interview with Supervising Public Health Sanitarian.

6. Interview with Associate Public Health Sanitarian, Buffalo, NY, April 22, 2008.

7. Americans with Disabilities Act, 42 U.S.C. §§ 12181–89 (2006).

8. Bruno Latour, "Where Are the Missing Masses? The Sociology of a Few Mundane Artifacts," in *Shaping Technology/Building Society: Studies in Sociotechnical Change,* ed. Wiebe E. Bijker and John Law (Cambridge, MA: MIT Press, 1992).

9. Buffalo's Chief Plumbing Inspector, Buffalo, NY, April 11, 2008.

10. See American National Standards Institute, "ANSI Standards Store," http://webstore.ansi.org (accessed July 20, 2008).

11. See United Spinal Association, "ANSI Endorses 'Visitability' Criteria," http://www.unitedspinal.org/publications/action/2008/04/08/ansi-endorses-"visitability"-criteria/ (accessed July 8, 2008).

12. Founded in 1880 as the American Society of Mechanical Engineers, ASME "promotes the art, science & practice of mechanical & multidisciplinary engineering and allied sciences around the globe. . . . [Its vision is to] [d]evelop the preeminent, universally applicable codes, standards, conformity assessment programs, and related products and services for the benefit of humanity. Involved the best and brightest people from around the world to develop, maintain, promote, and employ ASME products and services globally." See ASME International, "About Codes and Standards," http://www.asme.org/Codes/About/ (accessed July 8, 2008).

13. Plumbing Code of New York State, § 401.2. This standard "establishes requirements and test methods pertaining to materials, significant dimensions, and functional performances for vitreous china plumbing fixtures. The sanitary performance requirements and test procedures apply to all types of water closets and urinals that discharge into gravity waste systems in permanent buildings and structures, independent of occupancy. Features referenced in this Standard include water closets, lavatories, urinals, bidets, service sinks, drinking fountains, and institutional application fixtures." Product description for digital book, "A112.19.2–2003 Vitreous China Plumbing Fixtures and Hydraulic Requirements for Water Closets and Urinals," ASME website, http://catalog.asme.org/Codes/PrintBook/A112192_2003_Vitreous_China.cfm.

14. Plumbing Code of New York State, § 405.4.3.

15. Ibid., § 419.1.

16. Ibid., § 418.1.

17. Ibid., § 403.7.

18. Peter Jahrling, "Writing Restroom Specifications? Consider Traffic, Conservation, and Hygiene," *Construction Specifier*, March 2004.

19. Ibid. However, a *New York Times* article states that "after spending $5 million on its five automated public toilets, Seattle is calling it quits. In the end, the restrooms, installed in early 2004, had become so filthy, so overrun with drug abusers and prostitutes, that although use was free of charge, even some of the city's most destitute people refused to step inside them." See Christopher Maag, "Seattle to Remove Automated Toilets," *New York Times,* July 17, 2008.

20. John Fultz, "Get with the Program," *Reeves Journal,* June 28, 2003.

21. Unlike the flushometer, which embodies a gaze that is present only in the space of the washroom itself, the central computer manages the washroom from a central location elsewhere. Hence, the flushing device is not only programmed initially by the manufacturer but also involves continuous programming and reprogramming.

22. Fultz, "Get with the Program."

23. Ibid.

24. Ibid.

25. Newton Distributing, "Sloan ECOS (RESS-C)," http://www.newtondistributing.com/product/Sloan-ECOS-RESSC-489.html (accessed July 22, 2008).

26. Interview with Field Service Engineer, Sloan Valve, July 24, 2008.

27. Sloan's Field Service Engineer further notes that this ten-second rule does not work so well for some juveniles, who take less time to urinate and are thus illegible to the flushometer's infrared eye.

28. The data in this paragraph is based on "Beijing's Toilet Horrors Flushed Away as Olympics Near," posted on the *China Economic* website on June 21, 2008. See http://en.ce.cn/National/Local/200806/21/t20080621_15910533.shtml (accessed December 18, 2008).

29. See also Jennifer 8. Lee, "Lather Up: The Hand-Wringing over Hand-Washing," *New York Times,* February 22, 2008; and a *New Yorker* cartoon by Carolita Johnson, published July 26, 2006, depicting two hand-washing signs.

30. See, e.g., for schools, National Food Service Management Institute, "Wash Your Hands," http://www.olemiss.edu/depts/nfsmi/Information/handsindex.html (accessed July 22, 2008); and presentation video at http://www.ehow.com/video_12779_awash-hands.html (accessed July 23, 2008).

31. Delta Faucet Company, "Electronic Faucets," http://www.deltafaucet.com/wps/portal/deltacom/BathFeatures/ElectronicFaucets (accessed July 22, 2008).

32. Interview with Supervising Public Health Sanitarian.

33. For further illustration, see Nova Scotia Maple Syrup Company, "All Employees Must Wash Their Hands," http://www.novascotiamaplesyrup.com/Publications/HandWash.pdf (accessed July 26, 2008).

34. Interview with Field Service Engineer.

35. Foucault, *Discipline and Punish,* 178.

36. See PlumbingSupply.com, "Delta Electronic Faucets," http://www.plumbingsupply.com/delta_handsfree_faucets.html (accessed December 18, 2008).

37. New Home Source, "Selecting the Right Faucet," http://www.newhomesource.com/HomeGuideArticle/Article-deltaselectfaucet (accessed December 18, 2008). In addition, state regulations restrict the water temperature used in public washroom facilities. In a conversation in May 2007, an owner of a Buffalo gas station told me that his facility was shut down and he was required to pay large fees for inadequate water temperature in his washroom faucets.

38. Jahrling, "Writing Restroom Specifications?"

39. Lynn Marshall, "Many Say Toilets Have Got to Go," *Los Angeles Times,* September 9, 2007, http://articles.latimes.com/2007/sep/09/nation/na-toilets9 (accessed December 18, 2008).

40. It is not clear why there is such cost variation. William Saletan, "Crap and Trade: Why Public Toilets Should Pay You," *Slate,* July 9, 2008, http://www.slate.com/id/2195071/.

41. Michael Wilson, "Greetings, Earthlings. Your New Restroom Is Ready," *New York Times,* January 11, 2008.

42. Ibid.

43. Jennifer 8. Lee, "A Toilet That Uses 14 Gallons? 'Oh Gosh!'" *New York Times,* January 11, 2008.

44. "I Hate You, Fully Automated Bathroom," Craigslist, http://www.craigslist.org/about/best/sea/199973342.html (accessed July 11, 2008).

45. Matthew Crawford, *Shop Class as Soulcraft* (New York: Penguin, 2009), 55–56.

46. This was the 1970s slogan of environmentalist California governor Jerry Brown and an even older camping/southwestern adage. See Jack Taylor, "Water Conservation," http://www.flat-tax-and-term-limits.net/Water-Conservation-if-it's-yellow-let-it-mellow-if-it's-brown-flush-it-down.html (accessed December 27, 2008).

47. Telephone interview, December 23, 2008.

48. This last phrase brings to mind Jane Jacobs's approach toward city planning. See especially Jane Jacobs, *The Life and Death of Great American Cities* (New York: Random House, 1961). Here, instead of eyes on the street as providing safety, one might humorously suggest that it is "eyes on the slits" that do this work. My thanks to Harvey Molotch for suggesting this point.

49. This term was used by cultural geographer Sarah Whatmore in *Hybrid Geographies: Natures, Cultures, Spaces* (London: Sage, 2002).

50. See Michael Pollan, *The Botany of Desire: A Plant's Eye View of the World* (New York: Random House, 2001).

51. Latour, "Where Are the Missing Masses?"

52. Interview with Field Service Engineer.

53. Ibid.

54. Foucault, *Discipline and Punish*, 201.

Notes to Chapter 5

I would like to thank Harvey Molotch for his excellent and much needed guidance on development of this chapter. I am also grateful for the comments of the NYLON 2009 spring conference attendees.

1. The Bathroom Diaries is a website aimed at making tourism easier by providing a global, searchable database of public bathrooms: http://www.thebathroomdiaries.com/ (accessed February 20, 2009).

2. H. Marcovici, "New York City Cab Drivers and the Bathroom Dilemma," senior seminar paper, New York University, May 5, 2008, 14.

3. Schaller Consulting, *2006 Taxicab Handbook* (New York: B. Schaller, 2006).

4. Schaller Consulting, *The Changing Face of Taxi and Limousine Drivers: U.S., Large States and Metro Areas and New York City* (New York: B. Schaller, 2004).

5. Legal Action Center for the Homeless, *New York City's Public Restroom Crisis* (New York: Legal Action Center for the Homeless, October 1990), http://graphics8.nytimes.com/packages/pdf/nyregion/city_room/20081215_TOILETS.pdf; B. Mathis-Lilley, "Dead Heads," *New York*, March 6, 2006, http://nymag.com/news/intelligencer/16393/. See also "Rest Stop: Times Square Control," in this volume.

6. New York City Department of Parks and Recreation, *Parks and Recreation 2002–2003 Biennial Report: Eight Seasons of Progress*, Rebuilding Neighborhood Parks section.

7. M. Duneier, *Sidewalk* (New York: Farrar, Straus and Giroux, 2000).

8. M. Plaut, *Hack: How I Stopped Worrying about What to Do with My Life and Started Driving a Yellow Cab* (New York: Random House, 2007), 85 (emphasis in original).

9. P. Goldberger, "A Taxi Is Not a Car," in *Designing the Taxi*, ed. M. Canning, S. Gorton, and D. Marton (New York: Design Trust for Public Space, 2005), 6.

10. The $150 figure includes the lease fee and the cost of gas, which is the drivers' responsibility and was calculated from an average based on my interviewees' responses. Schaller Consulting, *2006 Taxicab Handbook*.

11. G. R. G. Hodges, *Taxi! A Social History of the New York City Cabdriver* (Baltimore: Johns Hopkins University Press, 2007).

12. Schaller Consulting, *2006 Taxicab Handbook*, 3.

13. Design Trust for Public Space, *Taxi 07: Roads Forward,* ed. R. Abrams, S. Elson, and C. Mauldin (New York: Design Trust for Public Space and the City of New York, 2007).

14. Schaller Consulting, *2006 Taxicab Handbook,* 3. Also see Schaller Consulting, *The Changing Face of Taxi and Limousine Drivers: U.S., Large States and Metro Areas and New York City* (New York: B. Schaller, 2004). The immigrant population in New York City had a sex ratio of ninety males to one hundred females in 2000 (City of New York, Department of City Planning, 2004).

15. D. Shoup, *The High Cost of Free Parking* (Chicago: American Planning Association, 2005).

16. M. Grace, "On East 78th Street, Neighbors Tell Taxis, No More Relief!" *New York Observer,* October 10, 2005.

17. D. Marton, "About Designing the Taxi," in *Designing the Taxi,* ed. M. Canning, S. Gorton, and D. Marton (New York: Design Trust for Public Space, 2005), 5.

18. B. Mathew, *Taxi! Cabs and Capitalism in New York City* (New York: New Press, 2005), 122.

19. Ibid., 116.

20. Three of my interviewees specifically mentioned fare beaters, people who jump out without paying. Seven of my interviewees talked about the difficulty of not always knowing how to get where a customer wants to go, especially in the outer boroughs and sometimes even within Manhattan. One driver had two cell phones, one of which was reserved specifically for accessing directions via the Internet.

21. M. Grynbaum, "Cabbies Stay on Their Phones Despite Ban," *New York Times,* August 3, 2009, available online at http://www.nytimes.com/2009/08/04/nyregion/04taxi.html?_r=1.

22. R. Sennett and J. Cobb, *The Hidden Injuries of Class* (New York: Vintage Books, 1972).

23. E. M. Dupuis, *Nature's Perfect Food: How Milk Became America's Drink* (New York: New York University Press, 2002).

24. K. Ascher, *The Works: Anatomy of a City* (New York: Penguin, 2005).

25. Dog owners are prevented from walking their dogs on park grass in those parks where dogs are prohibited.

26. Metropolitan Transit Authority, "Rules of Conduct," http://www.mta.info/nyct/rules/rules.htm.

27. New York City Department of Parks and Recreation, "Rules and Regulations," http://www.nycgovparks.org/sub_about/rules_and_regulations/rr_rules_regulations.html.

28. J. D. Goodman, "Scoop It Up or Pay: On Patrol with Enforcers of the Dog Law," *New York Times,* June 5, 2008, http://www.nytimes.com/2008/06/05/nyregion/05scooper.html. The Sanitation Department's Canine Task Force also writes tickets for off-leash dogs and for throwing household garbage into public trash receptacles.

29. M. Scorsese, director, and P. Schrader, writer, *Taxi Driver* (Bill/Philips, Columbia Pictures Corporation, and Italo-Judeo Productions, producers, 1976).

30. A. Kira, *The Bathroom,* 2d ed. (New York: Viking, 1976), 95.

31. Anatomically, though *E. coli* can live in the urethra, urethral infections are not considered serious enough to treat. It is only when the infection reaches the bladder via the urethra

that a bladder infection is caused. Women's urethras are shorter than men's—this is one proposed explanation for women's increased susceptibility to bladder infections. National Kidney and Urologic Disease Information Clearinghouse, National Institute of Health, "Urinary Tract Infection in Adults," http://kidney.niddk.nih.gov/Kudiseases/pubs/utiadult/.

32. J. Snow, *Mode of Communication of Cholera*, 2d ed. (London: John Churchill, New Burlington Street, 1855).

33. J. Brandow, *New York's Poop Scoop Law: Dogs, the Dirt, and Due Process* (West Lafayette, IN: Purdue University Press, 2008), 9.

34. A. Harivandi, "Lawns 'n' Dogs," Division of Agriculture and Natural Resources, University of California, Publication 8255 (2007).

35. Battery Park City Parks Conservancy, "The Impact of Dog Urine and Feces" (2006) (map of Battery Park and accompanying public information).

36. Plaut, *Hack*, 84.

37. Mayo Clinic, "Urinary Tract Infections," http://www.mayoclinic.com/health/urinary-tract-infection/DS00286 (2009).

38. National Kidney and Urologic Disease Information Clearinghouse, "Urinary Tract Infection in Adults."

39. Ascher, *The Works*, 140.

40. In a 2009 citywide call for taxi design suggestions put out by the Taxi and Limousine Commission, commenter Ivan P. wrote, "As you can see 95% of taxis [*sic*] drivers came from an undisciplined country. They came here with their selfish attitude and some mental problems." "Taxis of the Future," *Brian Lehrer Show*, WNYC, 2009, http://www.wnyc.org/shows/bl/episodes/2009/04/20/segments/129164.

41. N. Elias, *The History of Manners: The Civilizing Process*, vol. 1., trans. E. Jephcott (New York: Urizen Books, 1978).

42. Ibid., 178.

43. M. Foucault, *Discipline and Punish: The Birth of the Prison* (New York: Vintage, 1979).

44. Elias, *History of Manners*, 133; E. Goffman, *Relations in Public: Microstudies of the Public Order* (New York: Basic Books, 1971).

45. S. Strom, "Helmsley Left Dogs Billions in Her Will," *New York Times*, July 2, 2008.

46. Duneier, *Sidewalk*, 176.

47. G. G. Modan, *Turf Wars: Discourse, Diversity, and the Politics of Place* (Malden, MA: Blackwell, 2006), 106.

48. Kira, *The Bathroom*, 12, quoting Reginald Reynolds, *Cleanliness and Godliness* (Garden City, NY: Doubleday, 1946), 96.

Notes to Chapter 6

1. Lewis Mumford, *The City in History* (1935; repr., Harmondsworth, UK: Penguin, 1965).

2. Sue Cavanagh and Von Ware, *At Women's Convenience: A Handbook on the Design of Women's Public Toilets* (London: Women's Design Service, 1991).

3. *Architects Journal* (1953): 117, quoted in Cavanagh and Ware, *At Women's Convenience*, 9.

4. Alexander Kira, *The Bathroom* (Harmondsworth, UK: Penguin and Cornell University Press, 1976). Yoshiharu Asano, *Number of Fixtures: Mathematical Models* (Nagano, Japan: Faculty of Architecture and Building Engineering, Shinshu University, 2002); Asiah Rahim

(committee chair), *Public Toilets: Part 1—Minimum Design Criteria* (Kuala Lumpur: Department of Malaysian Standards, ICS: 03.080.30, 2006).

5. David Inglis, "Dirt and Denigration: The Faecal Imagery and Rhetorics of Abuse," in "The Toilet Issue," special issue of *Post Colonial Studies* 5, no. 2 (July 2002); Joanne Bichard, J. Hanson, C. Greed, et al., "Please Wash Your Hands," *Senses and Society* 2, no. 3 (2008): 385–90.

6. British Toilet Association, *Better Public Toilets: A Providers' Guide to the Provision and Management of "Away from Home" Toilets,* ed. Ray Fowler (Winchester, UK: British Toilet Association, 2001).

7. Steve Robinson, *Public Conveniences: Policy, Planning and Provision* (London: Institute of Wastes Management, 2001).

8. Clara Greed, *Inclusive Urban Design: Public Toilets* (Oxford, UK: Architectural Press, 2003), 91.

9. Judith Walkowitz, *City of Dreadful Delight: Narratives of Sexual Danger in Late-Victorian London* (Chicago: University of Chicago Press, 1992); Barbara Penner, "A World of Unmentionable Suffering: Women's Public Conveniences in Victorian London," *Journal of Design History* 14, no. 1 (2001): 35–54; Greed, *Inclusive Urban Design.*

10. Clara Greed, *Women and Planning: Creating Gendered Realities* (London: Routledge, 1994), 79–81.

11. Loretta Lees, ed., *The Emancipatory City? Paradoxes and Possibilities* (London: Sage, 2004). See also Walkowitz, *City of Dreadful Delight.*

12. Joanne Bichard et al., *Access to the Built Environment: Barriers, Chains and Missing Links* (London: Bartlett School of Graduate Architectural Studies, University College, London, 2004).

13. Clara Greed, *Surveying Sisters: Women in a Traditional Male Profession* (London: Routledge, 1991).

14. Ann De Graft-Johnson, S. Manley, and Clara Greed, *Why Do Women Leave Architecture?* (London: Royal Institution of British Architects, 2003).

15. Clara Greed and Isobel Daniels, *User and Provider Perspectives on Public Toilet Provision* (Nuffield Trust funded research, Occasional Paper, University of the West of England, Bristol, 2002).

16. Clara Greed, "Overcoming the Factors Inhibiting the Mainstreaming of Gender into Spatial Planning Policy in the United Kingdom," *Urban Studies* 42, no. 4 (2005): 1–31; Clara Greed, "An Investigation of the Effectiveness of Gender Mainstreaming as a Means of Integrating the Needs of Women and Men into Spatial Planning in the United Kingdom," *Progress in Planning* 64, no. 4 (2005): 239–321.

17. Dolores Hayden, *Redesigning the American Dream* (New York: Norton, 2002); James Kunstler, *The Geography of Nowhere: The Rise and Decline of America's Man-Made Landscape* (New York: Touchstone, 1994).

18. In this respect, see the competition for high-concept public toilets developed by the Royal Institute of British Architects (RIBA).

19. Marion Roberts and Clara Greed, *Approaching Urban Design: The Design Process* (Harlow, UK: Longmans, 2001); Peter Jones, Marion Roberts, and Linda Morris, *Rediscovering Mixed-Use Streets: The Contribution of Local High Streets to Sustainable Communities* (Bristol, UK: Policy, 2007).

20. Greed and Daniels, *User and Provider Perspectives.*

21. See NewUrbanism.org, "Principles," http://www.newurbanism.org/newurbanism/principles.html.

22. Susan Cunningham and Christine Norton, *Public Inconveniences: Suggestions for Improvements* (London: All Mod Cons and the Continence Foundation, 1993).

23. Yutaka Miyanishi, *Comfortable Public Toilets: Design and Maintenance Manual* (Toyama, Japan: City Planning Department, 1996).

Notes to Chapter 7

1. See, e.g., discussion later in this chapter about separate toilets in ladies' reading rooms in mid-nineteenth-century public libraries.

2. Act of Mar. 24, 1887, ch. 103, § 2, 1887 Mass. Acts 668.

3. George Martin Kober, "History of Industrial Hygiene and Its Effects on Public Health," in *A Half Century of Public Health*, ed. Mazÿck P. Ravenal (New York: American Public Health Association, 1921), 377.

4. Mary P. Ryan, *Women in Public: Between Banners and Ballots, 1825–1880* (Baltimore: Johns Hopkins University Press, 1990), 64.

5. Catherine Clinton, *The Other Civil War* (New York: Hill and Wang, 1984), 18.

6. The separation of the worlds of the household and the workplace was not a sudden event but rather a "slow and tangled disengagement." Ibid., 23.

7. See, e.g., Catherine Clinton and Christine Lunardini, *The Columbia Guide to American Women in the Nineteenth Century* (New York: Columbia University Press, 2000), 36–37.

8. David E. Shi, *Facing Facts: Realism in American Thought and Culture, 1850–1920* (New York: Oxford University Press, 1995), 17.

9. Barbara Welter, *Dimity Convictions: The American Woman in the Nineteenth Century* (Athens: Ohio University Press, 1976), 21.

10. As early as 1822, when textile mills were founded in Lowell, Massachusetts, young women flocked to mill towns, and white single women constituted the overwhelming majority of the early textile work force. Clinton, *Other Civil War*, 22.

11. Ryan, *Women in Public*, 31–32; Clinton, *Other Civil War*, 54.

12. Ryan, *Women in Public*, 79.

13. Ibid., 76.

14. Cynthia Eagle Russett, *Sexual Science: The Victorian Construction of Womanhood* (Cambridge, MA: Harvard University Press, 1989), 10.

15. See Suellen Hoy, *Chasing Dirt* (Oxford: Oxford University Press, 1995), 3.

16. Ibid., 18.

17. Ibid., 23. See also Maureen Ogle, *All the Modern Conveniences: American Household Plumbing, 1840–1890* (Baltimore: Johns Hopkins University Press, 1996), 106: "Physicians and laypeople alike . . . accepted the doctrine of predisposing causes—that is, the idea that people's behavior rendered them susceptible to disease."

18. Ogle, *Modern Conveniences*, 102; See generally John Duffy, *The Sanitarians: A History of American Public Health* (Urbana: University of Illinois Press, 1990).

19. John F. Kasson, *Rudeness and Civility: Manners in Nineteenth-Century Urban America* (New York: Hill and Wang, 1990), 116.

20. See ibid., 124: "In public especially, but also in private, one sought particularly to stifle all activities that might draw attention to the internal workings of the body, such as

coughing, sneezing, yawning, scratching, tooth picking, throat clearing, and nose blowing. More intimate functions were generally beneath discussion." In an admonition against chewing tobacco, Kasson quotes one etiquette writer's view that spit "is an excrement of the body, and should be disposed of as privately and carefully as any other." Ibid., 126.

21. Elizabeth Wilson, *The Sphinx in the City: Urban Life, The Control of Disorder, and Women* (Berkeley: University of California Press, 1991), 37. Addressing the deplorable health conditions in London around midcentury, Wilson notes,

> Efficient sewage systems were as desperately needed as adequate water supplies. In both cases, morality was inextricably entwined with cleanliness, disorder with filth. For the Victorians excrement became a metaphor, and a symbol for moral filth, perhaps even for the working class itself, and when they spoke and wrote of the cleansing of the city of filth, refuse and dung, they may really have longed to rid the cities of the labouring poor altogether.

22. Shi, *Facing Facts*, 13.

23. Ibid., quoting novelist Hjalmar H. Boyeson, "The Realism of American Fiction," *Independent* 44 (1892): 3. The nation's cultural elite focused on refining tastes, culture, and spiritual sensibilities and "took little interest in the commonplaces of everyday life." Shi, *Facing Facts*, 15, 17.

24. Shi, *Facing Facts*, 3–5.

25. See, e.g., Duncan Kennedy, "Toward an Historical Understanding of Legal Consciousness: The Case of Classical Legal Thought in America, 1850–1940," *Research in Law and Sociology* 3 (1980): 23; Donald H. Gjerdingen, "The Future of Our Past: The Legal Mind and the Legacy of Classical Common-Law Thought," *Indiana Law Journal* 68 (1993): 743; William M. Wiecek, *The Lost World of Classical Legal Thought: Law and Ideology in America, 1886–1937* (New York: Oxford University Press, 1998), 79–93.

26. Donald H. Gjerdingen, "The Politics of the Coase Theorem and Its Relationship to Modern Legal Thought," *Buffalo Law Review* 35 (1986): 877: "[P]eople have rights in things. Physical spaces and things are the usual objects of control; physical dominion is the usual manifestation of control. Physical boundaries thus become important and designate the things subject to dominion."

27. Shi, *Facing Facts*, 5.

28. Russett, *Sexual Science*, 3.

29. Ibid., 4.

30. Ibid., 11.

31. Ibid., 12.

32. Abigail A. Van Slyck, "The Lady and the Library Loafer: Gender and Public Space in Victorian America," *Winterthur Portfolio* 31 (1996): 223.

33. Ibid., 221.

34. Ibid., 227.

35. Ibid., 241.

36. Barbara Young Welke has explored how the growth of railroads and urban streetcars in the United States affected social understandings of gender and race during the second half of the nineteenth century. Barbara Young Welke, *Recasting American Liberty: Gender, Race, Law, and the Railroad Revolution, 1865–1920* (Cambridge: Cambridge University Press, 2001).

37. Ibid., 254.

38. See Katherine C. Grier, "Imaging the Parlor, 1830–1880," in *Perspectives on American Furniture*, ed. Gerald W. R. Ward (New York: Published for the Henry Francis du Pont Winterthur Museum by Norton, 1988), 205.

39. Ibid., 234–35.

40. Ibid., 239. Lynne Walker has suggested that late-nineteenth-century separate parlor spaces provided Victorian society with a way to deal with the growing presence of women in public while still adhering to the separate-spheres ideology: "These department stores provided a setting in the public sphere, 'a meeting place and promenade,' which for the first time gave women 'a home away from home,' a feeling of being at home in the public sphere, which only men had previously experienced." Lynne Walker, "Vistas of Pleasure: Women Consumers of Urban Space in the West End of London, 1850–1900," in *Women in the Victorian Art World*, ed. Clarissa Campbell Orr (Manchester: Manchester University Press, 1995), 79.

41. Ryan, *Women in Public*, 77–78.

42. See, e.g., Elizabeth Brandeis, "Labor Legislation," in *History of Labor in the United States, 1896–1932*, vol. 3, ed. John R. Commoner (1935; repr., Augustus M. Kelley, 1966), 461–62, Minimal statutory protections for all adult industrial workers without regard to sex did not come about in the United States until the late 1930s. See Alice Kessler-Harris, "The Paradox of Motherhood: Night Work Restrictions in the United States," in *Protecting Women: Labor Legislation in Europe, the United States, and Australia, 1880–1920*, ed. Ulla Wikander, Alice Kessler-Harris, and Jane Lewis (Urbana: University of Illinois, 1995), 338.

43. 40 Act of March 19, 1852, sec. 1, 1852 Ohio Laws 187.

44. The first health and safety legislation aimed at barring women from certain professions was an 1872 Illinois law forbidding women from working in mines. Act of Mar. 27, 1872, § 6, 1872 Ill. Laws 570 (providing for the health and safety of persons employed in coal mines).

45. See, e.g., Act of Apr. 11, 1890, ch. 183, § 1, 1890 Mass. Acts 152; 1889 N.Y. Sess. Laws ch. 560.

46. See, e.g., Act of Apr. 22, 1887, ch. 215, 1887 Mass. Acts 832.

47. See, e.g., Act of Mar. 31, 1915, Me. Rev. Stat. Ann., ch. 350, § 1 (1916).

48. See, e.g., Act of May 26, 1913, ch. 112, 1913 Conn. Pub. Acts 1701; Act of Apr. 15, 1912, ch. 331, sec. 1, § 93-a, 1912 N.Y. Laws 660.

49. See, e.g., Act of May 12, 1919, No. 239, § 1, 1919 Mich. Pub. Acts 427. (Caveat appended to wage nondiscrimination law: "Provided, however, That no female shall be given any task . . . detrimental to her morals, her health or her potential capacity for motherhood.").

50. See, e.g., Act of May 18, 1881, ch. 298, 1881 N.Y. Laws 402; Act of Apr. 12, 1882, ch. 150, 1882 Mass. Pub. Stats. Supp. 28; Act of Mar. 31, 1882, ch. 159, 1882 N.J. Laws 227; Act of Feb. 27, 1883, ch. 45, 1883 Neb. Gen. Laws 229.

51. See, e.g., 1919 Mich. Pub. Acts, 427 (provided that no woman "shall be given any task disproportionate to her morals, her health, or her potential capacity for motherhood").

52. Grace F. Ward, "Weakness of the Massachusetts Child Labor Laws," in Charles E. Persons, "The Early History of Factory Legislation in Massachusetts," in *Labor Laws and Their Enforcement*, ed. Charles E. Persons, Mabel Parton, and Mabelle Moses (1911; repr., New York: Arno, 1971), 161.

53. See, e.g., Robert Ritchie, *Observations on the Sanatory Arrangements of Factories with Remarks on the Present Methods of Warming and Ventilation and Proposals for Their*

Improvement (London: J. Weale, 1844) (study of British factories focusing almost entirely on ventilation and airborne poisons).

54. An 1882 study of sanitation in British factories states,

> General Evidences as to Sanitary Defects and Consequences. . . .
>
> Mr. R.W. Cole stated that in large factories they are in the habit of putting closets in the workroom, which sometimes are exceedingly injurious to the health. They emit the most frightful odour, so much so, that I have often felt sickened myself in going through the factories.

B. H. Thwaite, *Our Factories, Workshops, and Warehouses, Their Sanitary and Fire-Resisting Arrangements* (London: Spon, 1882), 26.

55. "Men's Ready-Made Clothing," in *Report on Condition of Woman and Child Wage-Earners in the United States in 19 Volumes* (1910), Senate Doc. 61-645 (prepared under the direction of Chas. P. Neill, Commissioner of Labor), vol. 3, 499, quoting "Second Annual Report of the Factory Inspectors of New York" (1887), 26.

56. George M. Price, *The Modern Factory: Safety, Sanitation and Welfare* (New York: Wiley, 1914), 275. Price notes that "all industrial and sanitary codes demands [*sic*] separate water-closet compartments for the sexes in every factory where men and women are employed. All toilet rooms should be located within the factory building and be convenient and accessible to the persons using them." Ibid., 277.

57. Some other states also adopted toilet-separation laws in gender-neutral language. See, e.g., 1913 N.C. Sess. Laws, ch. 83, § 1, 127 ("An Act to Compel All Persons and Corporations Engaged in Manufacturing or Other Business Enterprises Where Male and Female Employees Are Employed to Provide Separate and Distinct Toilets.").

58. Act of May 25, 1887, ch. 462, § 13, 1887 N.Y. Laws 575 ("A suitable and proper washroom and water-closets shall be provided for females where employed, and the water-closets used by females shall be separate and apart from those used by males.").

59. See, e.g., 1893 Pa. Laws, No. 244, 276; 1919 N.D. Laws, ch. 174, 317; 1913 S.D. Sess. Laws, ch. 240, 332; 1891 Ohio Laws, No. 413, 87; 1919 Ark. Acts 197; 1897 Tenn. Pub. Acts, ch. 98, § 1, 247.

60. By the early twentieth century, great interest developed in examining the working conditions in American factories. States established commissions to study factory health issues, while at the federal level the Department of Labor undertook investigations into factory sanitation. See, e.g., C. F. W. Doehring, "Factory Sanitation and Labor Protection," *Bulletin of the Department of Labor* 44 (1903): 1. In addition, in 1907, the U.S. Congress passed "An Act to authorize the Secretary of Commerce and Labor to investigate and report upon the industrial, social, moral, educational, and physical condition of woman and child workers in the United States." As a result, in 1910, the Department of Commerce and Labor issued an extended study entitled *Report on Condition of Woman and Child Wage-Earners in the United States in 19 Volumes*.

61. Doehring, "Factory Sanitation and Labor Protection," 1–2. Such concerns often focused on potential damage that factory work might cause a woman's reproductive capacity. See, e.g., ibid., 28 (section of report concerning "Relation of Sex to Lead Poisoning," quoting Dr. Thomas Oliver): "Where the two sexes are as far as possible equally exposed to the influence of lead, women probably suffer more rapidly, certainly more severely, than men. To a certain extent the reason is to be found in the fact that lead exercises an injurious influence upon the reproductive functions of women. It deranges menstruation."

In a 1908 monograph on occupational diseases, the same Dr. Oliver wondered whether the increased speed of machinery might result in female workers mothering "infants who are puny, ill-nourished and of a highly strung nervous system." Thomas Oliver, *Diseases of Occupation from the Legislative, Social and Medical Points of View* (New York: Dutton, 1908), 3.

62. George M. Price, Joint Board of Sanitary Control in the Dress and Waist Industry, *Special Report on Sanitary Conditions in the Shops of the Dress and Waist Industry* (New York: Schreiber, 1913), 13.

63. Ibid., 16.

64. Ogle, *Modern Conveniences*, 93. See also ibid., 111: "The enthusiasm with which Americans embraced sanitary science stands as clear evidence of the extent to which the cultures of scientism and professionalism had taken hold. During the middle of the century, Americans had paid little attention to the sanitary reformers among them, but in the late nineteenth century, they heeded the sanitarians' call and set about transforming their homes and their cities."

65. Among J. J. Cosgrove's publications were *History of Sanitation* (Pittsburgh: Standard Sanitary Manufacturing, 1909); *Principles and Practice of Plumbing* (Pittsburgh: Standard Sanitary Manufacturing, 1906); *Sewerage Purification and Disposal* (Pittsburgh: Standard Sanitary Manufacturing, 1909); *Wrought-Pipe Drainage Systems* (Pittsburgh: Standard Sanitary Manufacturing, 1909); *Plumbing Estimates and Contracts* (Pittsburgh: Standard Sanitary Manufacturing, 1910); *Plumbing Plans and Specifications* (Pittsburgh: Standard Sanitary Manufacturing, 1910); *Design of the Turkish Bath* (Pittsburgh: Standard Sanitary Manufacturing, 1913); *Factory Sanitation* (Pittsburgh: Standard Sanitary Manufacturing, 1913); and *Sanitary Refrigeration and Ice Making* (Pittsburgh: Standard Sanitary Manufacturing, 1914).

66. Cosgrove, *Plumbing Plans and Specifications*, 222–23.

67. This is well illustrated by George Price's *Special Report*. The following are among the recommendations set forth as "Sanitary Standards of the Industry":

20. Walls and ceiling of shops and water-closet apartments should be cleaned as often as necessary, and kept clean.

21. Floors of shops, and of water-closet apartments, to be scrubbed weekly, swept daily, and kept free of refuse.

22. A separate water-closet apartment shall be provided for each sex, with solid partitions to extend from floor to ceiling, and with separate vestibules and door.

23. Water-closets to be adequately flushed and kept clean.

. . .

27. Water-closet apartments, dressing-rooms, wash-rooms, and lunch-rooms to be properly lighted, ventilated, and cleaned, and to be kept clean.

Ibid., 16. Sex separation is melded with a range of other "sanitary standards." Another example of a sanitary expert's conflating sex separation with other sanitary requirements is Milton J. Rosenau, *Preventive Medicine and Hygiene* ([New York: D. Appleton, 1921], 1337):

Water-closets and Urinals.—Separate accommodations must be provided for the sexes; privies in country districts should be in entirely separate buildings. The urinals should be constantly and automatically flushed and water-closets and urinals should be made to allow complete inspection and use of the scrubbing brush. Thorough ventilation of the toilet-rooms should be planned for and they should be kept clean and sweet at all times.

68. Act of February 6, 1889, ch. 4, § 1, 1889 Cal. Laws 3.

69. 1921 Conn. Pub. Acts 3250.

70. See, e.g., 1899 Ind. Acts 235.

71. See, e.g., 1893 Pa. Laws 278.

72. 1921 Conn. Pub. Acts 3250.

73. Kasson, *Rudeness and Civility,* 124.

74. In the Department of Commerce and Labor report on the glass industry, in a section titled "Closets for Females," the reporter notes that the privacy of approach to factory toilets is a "factor affecting the character of women's closets . . . to which very few manufacturers have given any thought or, at least, active attention." He criticizes instances in which "closets are often simply walled-off portions of the workroom, the men's and women's closets side by side and the entrances exposed to the direct view of all." "Glass Industry," in *Report on Condition of Woman and Child Wage-Earners,* vol. 3, 353.

75. "Cotton Textile Industry," in *Report on Condition of Woman and Child Wage-Earners,* vol. 1, 371.

76. In the Department of Commerce and Labor's report on laundries, the reporter criticized the location of the water closet:

> In one instance the one closet of the establishment was in the basement under the pavement, with no light except what came from a circular piece of glass set in the pavement and no ventilation but the open door. . . . It was in plain sight of the men who were doing the washing in the basement, clad only in their undergarments. . . . In another laundry, while other conditions were very good, the one objectionable feature was one closet for both sexes and that in a dark, unwholesome basement, where the women had to pass among the washers to reach it.

"Laundries," in *Report on Condition of Woman and Child Wage-Earners,* vol. 12, 12. The concern is twofold: First, women entering the water closet might be seen by male workers. Second, women workers will have to endure the embarrassing sight of male workers in their underwear.

77. The literature surrounding factory sanitation abounds with examples of the conflation of concerns of morality with concerns of women's health and modesty. For example, 1913 Or. Laws 92, enacted "to protect the lives and health and morals of women and minor workers," stated, "It shall be unlawful to employ women or minors in any occupation . . . for unreasonably long hours; and it shall be unlawful to employ women or minors in any occupation . . . under such surroundings or conditions, sanitary or otherwise, as may be detrimental to their health or morals."

Elsewhere, a 1905 report on cordage and twine factories states,

> The evidence . . . indicates that the menstruation of the doffers, usually young girls between fourteen and seventeen years of age, is affected by their standing barefoot on wet floors, and that colds, rheumatism and bronchitis are unduly prevalent among wet-room women.
>
> Several doctors in twine and cordage factory towns have suggested that the relation between wet-room conditions and morals is as important a subject for investigation as the one chosen, and other indications bear out this opinion.

Mabel Parton, "The Work of Women and Children in Cordage and Twine Factories," in *Labor Laws and Their Enforcement,* ed. Susan M. Kingsbury (New York: Longmans, Green, 1911), 144–45. In a section of an investigation into the cotton textile industry entitled "The Moral Condition of Cotton-Mill Operatives," the writer states,

In cotton mills large numbers of men, women, and children are brought together in the same workrooms. Where men and women are thus constantly associated it is, of course, possible for immoral relations between them to spring up. A woman and a man never jointly attend the same machine; usually each has several large machines to attend. A woman, if she wishes, need have no communication with the men in the mill except the section hand, second hand, and overseer.

"Cotton Textile Industry," in *Report on Condition of Woman and Child Wage-Earners*, vol. 1, 590. The very next paragraph ties concerns of privacy directly to those of morality: "In many mills . . . there is no privacy of approach to the toilets, and anyone entering them does so in full view of persons of both sexes in the same workroom, a condition obviously not in the interest of good morals." Ibid.

78. Cosgrove, *Factory Sanitation*.

79. Thus, in a section entitled "Interesting Men in Their Employment," Cosgrove states, "Washing and bathing facilities will bring the men to work ahead of time so they can change their clothes; throws them together so they become better acquainted; makes a sort of social club of the factory outside of working hours, and promotes a good feeling all around." Ibid., viii.

80. Ibid., ix.

81. Ibid., xxii.

82. Thwaite, *Our Factories, Workshops, and Warehouses*, 36.

Notes to Chapter 8

Support for this essay was made possible by a grant from the Academic Senate of the University of California, San Diego. Thanks to David Benin and Lauren Berliner for research assistance; Val Hartouni, Robert McRuer, Joseph Masco, and Brian Selznick for superb comments on early drafts; and Harvey Molotch and Laura Norén for editorial persistence, guidance, and encouragement.

1. See Roberta Ann Johnson, "Mobilizing the Disabled" (1983), in *Waves of Protest: Social Movements Since the Sixties*, ed. Jo Freeman and Victoria Johnson (New York: Rowman and Littlefield, 1999), 25–45; see also Joseph P. Shapiro, *No Pity: People with Disabilities Forging a New Civil Rights Movement* (New York: Three Rivers, 1993).

2. Quoted in Doris Zames Fleischer and Freida Zames, *The Disability Rights Movement: From Charity to Confrontation* (Philadelphia: Temple University Press, 2001), 68.

3. Fleischer and Zames, *Disability Rights Movement*, 93–102.

4. Shapiro, *No Pity*, 67.

5. Carol A. Breckenridge and Candace Vogler, "The Critical Limits of Embodiment: Disability's Criticism," *Public Culture* 13 (2001): 350, emphasis mine.

6. Henri-Jacques Stiker, *A History of Disability*, trans. William Sayers (Ann Arbor: University of Michigan Press, 2000), 128.

7. One episode of the U.S. cable television series *Curb Your Enthusiasm* revolves around the lead character's use of a disabled stall in a men's public toilet in a way that inconveniences and comes to anger a disabled wheelchair user and raises provocative questions about the deeply gendered dimensions of what happens when men share a public toilet experience. See "The Bowtie" (originally broadcast October 2, 2005), *Curb Your Enthusiasm: The Complete Fifth Season*, DVD boxed set (Home Box Office, 2006).

8. See Allan Bérubé, "The History of Gay Bathhouses" (1984), in *Policing Public Sex: Queer Politics and the Future of AIDS Activism*, ed. Dangerous Bedfellows (Boston: South End, 1996), 187–220. See also David Serlin, "Bathhouses," in *The Encyclopedia of American Lesbian, Gay, Bisexual, and Transgender History in America*, ed. Marc Stein (New York: Scribner's, 2004), 122–25.

9. Selwyn Goldsmith, *Designing for the Disabled: The New Paradigm* (Oxford, UK: Architectural Press, 2001), 79. According to Goldsmith, the British design equivalent to the A117.1, the BS5810, was developed to be seventy-nine inches deep, suggesting a generosity of spatial accommodation meant to impart a generosity of spirit after decades of implicit and often explicit spatial segregation (ibid., 183).

10. Anecdote related by Katherine Ott during interview with author, July 2005. See David Serlin, "Making History Public: An Interview with Katherine Ott," *Radical History Review* 94 (Winter 2006): 197–211.

11. For a further exploration of the history of disability rights activism, see James I. Charlton, *Nothing about Us without Us: Disability Oppression and Empowerment* (Berkeley: University of California Press, 2000).

12. See, for example, David Serlin, "Disability, Masculinity, and the Prosthetics of War, 1945 to 2005," in *The Prosthetic Impulse: From a Posthuman Present to a Biocultural Future*, ed. Marquard Smith and Joanne Morra (Cambridge, MA: MIT Press, 2006), 155–83.

13. See Susan Schweik, *The Ugly Laws* (New York: New York University Press, 2009).

14. See Maureen Ogle, *All the Modern Conveniences: American Household Plumbing, 1840–1890* (Baltimore: Johns Hopkins University Press, 2000).

15. See, for example, Julian B. Carter, *The Heart of Whiteness: Normal Sexuality and Race in America, 1880–1940* (Durham, NC: Duke University Press, 2007); Sarah E. Igo, *The Averaged American: Surveys, Citizens, and the Making of a Mass Public* (Cambridge, MA: Harvard University Press, 2008); and Jennifer Terry, *An American Obsession: Science, Medicine, and Homosexuality in Modern Society* (Chicago: University of Chicago Press, 1999).

16. Noah McClain, "Making Messes: Fare-Beating in the Subway, Instruments of Control, and the Trajectories of Control," unpublished paper, Department of Sociology, New York University, 2009.

17. Goldsmith, *Designing for the Disabled*, 180–83.

18. Eva Feder Kittay, *Love's Labor: Essays on Women, Equality, and Dependency* (New York: Routledge, 1999).

19. Ibid., 93.

20. See David Serlin, *Replaceable You: Engineering the Body in Postwar America* (Chicago: University of Chicago Press, 2004), esp. 39–48.

21. Kittay, *Love's Labor*, 171–72.

22. Sue Bettison, *Toilet Training to Independence for the Handicapped: A Manual for Trainers* (Springfield, IL: Charles C. Thomas, 1982), 24.

23. Daniel J. Wilson, "Fighting Polio like a Man: Intersections of Masculinity, Disability, and Aging," in *Gendering Disability*, ed. Bonnie G. Smith and Beth Hutchison (New Brunswick, NJ: Rutgers University Press, 2004), 122.

24. Nancy Mairs, *Waist-High in the World: A Life among the Nondisabled* (Boston: Beacon, 1996), 96.

25. Jennifer Levi and Bennett Klein, "Pursuing Protection for Transgender People through Disability Laws," in *Transgender Rights*, ed. Paisley Currah, Richard M. Juang, and Shannon Price Minter (Minneapolis: University of Minnesota Press, 2006), 77.

26. See Patrick White, "How the Blind Became Heterosexual," *GLQ: A Journal of Lesbian and Gay Studies* 9, nos. 1–2 (2003): 133–48.

27. Erving Goffman, "The Arrangement between the Sexes," *Theory and Society* 4, no. 3 (September 1977): 316, emphasis in original.

28. For a good example of contemporary genderqueer activism around toilets, see Simone Chess, Alison Kafer, Jessi Quizar, and Mattie Udora Richardson, "Calling All Restroom Revolutionaries!" in *That's Revolting! Queer Strategies for Resisting Assimilation*, ed. Matt Bernstein Sycamore (Brooklyn, NY: Soft Skull, 2004), 189–206.

29. See, for example, Goldsmith, *Designing for the Disabled*.

30. Raymond Lifchez and Barbara Winslow, *Design for Independent Living: The Environment and Physically Disabled People* (Berkeley: University of California Press, 1980), 90.

31. Patrick Joyce, *The Rule of Freedom: Liberalism and the Modern City* (New York: Verso, 2004), 98. Joyce's reading of urban behavior in cities such as London, Manchester, Prague, and Vienna becomes part of a Foucauldian reclamation of the controlling machinations of urban life, tempering and ultimately revising several decades' worth of scholarship on the *flâneur*, the mobile, disinterested urban spectator who neither possesses nor exhibits any modicum of allegiance to the terrains through which he or she traverses. For a comparative approach, see also Christine M. Boyer, *Dreaming the Rational City: The Myth of American City Planning* (Cambridge, MA: MIT Press, 1983).

32. For more about Holdsworth, see Mike Ervin, "Johnny Crescendo, British Balladeer for Disability Rights," *Disability World* 14 (June–August 2002), http://www.disabilityworld.org/06-08_02/il/crescendo.shtml (accessed January 8, 2009).

33. For a superb overview of the politics of the medical-telethon-entertainment complex, see Paul Longmore, "Conspicuous Contribution and American Cultural Dilemmas: Telethon Rituals of Cleansing and Renewal," in *The Body and Physical Difference: Discourses of Disability*, ed. David T. Mitchell and Sharon L. Snyder (Ann Arbor: University of Michigan Press, 1997), 134–58.

Notes to Chapter 9

1. Barbara Penner, "A World of Unmentionable Suffering: Women's Public Conveniences in Victorian London," *Journal of Design* 14, no. 2 (2001): 35–51.

2. John H. Ohly, "History of Plant Seizures during World War II," Office of the Chief of Military History, Department of the Army, vol. 3, 1946, appendix Z-1-a.

3. Emily Yellin, *Our Mothers' War* (New York: Free Press, 2005).

4. See Leslie Feinberg, *Transgender Warriors: Making History from Joan of Arc to Dennis Rodman* (Boston: Beacon, 1996); Kate Bornstein, *My Gender Workbook* (London and New York: Routledge, 1998); Dylan Vade, "Gender Neutral Bathroom Survey," unpublished paper, Transgender Law Center, San Francisco, CA, 2005.

5. Patricia Cooper and Ruth Oldenziel, "Cherished Classifications: Bathrooms and the Construction of Gender/Race on the Pennsylvania Railroad during World War II," *Feminist Studies* 25, no. 1 (1999): 7.

6. Ibid., 8.

7. A. J. Crittendon, Rori Hanson, Cole Popp, Dylan Larke, and Andrew Schiffer, "Gender Neutral Bathroom Facilities in the Residence Halls," memorandum to the University of Massachusetts administration, December 14, 2001.

8. Ibid.

9. Amanda M. Dove, "Mission," email to the Restroom Revolution members, November 14, 2001.

10. Mitch Boucher et al., "Gender Neutral Bathrooms on Campus," memorandum to the University of Massachusetts administration, October 2, 2002.

11. Ibid.

12. Benjamin Gedan, "Group Wants Transgender Bathrooms for UMass," *Boston Globe,* October 20, 2002.

13. Lucio Guerrero, "U. of C. Activists Want Bathrooms without Gender," *Chicago Sun-Times,* November 26, 2003.

14. Dylan Vade, "Bathrooms," email to Mitch Boucher, May 31, 2002.

15. William Weir, "Gender Won't Count in the New Dorm," *Hartford Courant,* May 18, 2003.

16. Gedan, "Group Wants Transgender Bathrooms."

17. Dan Lamothe, "Restroom Revolution: A Big Step," *Daily Collegian,* November 26, 2002.

18. Courtney Charles, "The Restroom Revolution: Lobbying for Lavatory Change," *Daily Collegian,* December 10, 2002.

19. Mark Hoffman, posting on the *Daily Collegian* electronic bulletin board, October 4, 2002, available at http://www.dailycollegian.com.

20. Olaf Aprans, "Transgenderism," *Minuteman,* November 26, 2002.

21. "The Politics of Pee," *Minuteman,* November, 26, 2002.

22. Louis P. Sheldon, "A Gender Identity Disorder Goes Mainstream," Traditional Values Coalition, October 24, 2002, available at http://traditionalvalues.org.

23. Steve Miller, "Down the Drain," *Independent Gay Forum,* October 22, 2002, available at www.indegayforum.org.

24. Aprans, "Transgenderism."

25. Cooper and Oldenziel, "Cherished Classifications," 17.

26. Janie Kim, "Do You Feel That 'Transgendered' Students Should Have Their Own Bathrooms, Taking away a Men's or Women's Room from Buildings?" *Minuteman,* November 26, 2002.

27. Ibid.

28. Peter C., posting on the *Daily Collegian* electronic bulletin board, October 3, 2002, available at http://www.dailycollegian.com.

29. Jason Pierce, posting on the *Daily Collegian* electronic bulletin board, October 6, 2002, available at http://www.dailycollegian.com.

30. Adam D., posting on the *Daily Collegian* electronic bulletin board, October 6, 2002, available at http://www.dailycollegian.com.

31. Louise Antony, "Back to Androgyny: What Bathrooms Can Teach Us about Equality," *Journal of Contemporary Legal Issues* 9 (1998): 9.

32. Lamothe, "Restroom Revolution."

33. Jane S. De Hart, "Gender on the Right: Meanings behind the Existential Scream," *Gender and History* 3 (1991): 255–56.

34. Mitch Boucher, "The Truth about the Restroom Revolution," *Daily Collegian,* December 4, 2002.

35. Kristin Shrewsbury, "SGA Votes on New Restrooms," *Daily Collegian,* November 14, 2002.

36. Ibid.

37. Kiera Manikoff, "A Tale of Two Genders . . . and Then Some?" *Minuteman*, November 26, 2002.

38. Jared Nokes, letter to the editor, *Daily Collegian*, December 2, 2002.

39. "Jackass of the Month: Ed Kammerer," *Minuteman*, November 26, 2002.

40. See *Jette v. Honey Farms Mini Market*, No. 95 SEM 0421 (MCAD, October 10, 2001); *Millet v. Lutco*, 23 M.D.L.R. 231 (2001); *Doe v. Yunits*, No. 00-1060-A (Mass. Super. Ct., October 11, 2000).

41. Richard A. Wasserstrom, "Racism, Sexism, and Preferential Treatment: An Approach to the Topics," *UCLA Law Review* 24 (1977): 581–615.

42. Ibid., 610.

43. Ibid., 606.

44. Ibid., 591–92.

45. Ibid., 594.

46. Antony, "Back to Androgyny," 10.

47. Ibid., 11.

Notes to Chapter 10

1. Jane J. Mansbridge, *Why We Lost the ERA* (Chicago: University of Chicago Press, 1986), 113.

2. The quoted language is taken from Sec. 61.43 of the Texas Alcoholic Beverage Code, but it is typical of that found in a large number of other codes.

3. Wendy Shalit, "A Ladies' Room of One's Own," *Commentary*, August 1995; Wendy Shalit, *A Return to Modesty: Discovering a Lost Virtue* (New York: Simon and Schuster, 1999).

4. See, e.g., "'Ally,' Access Laws Make Stalls for All, Businesses Follow Unisex Toilet Trend," *USA Today*, May 4, 1998, 1D.

5. See "Feminist Groups Can't Choose Bathrooms," *Rush Limbaugh Show*, November 26, 2003, available at http://mfile.akamai.com/5020/wma/rushlimb.download.akamai.com/5020/clips/03/11/112603_6_feminists.asx.

6. Associated Press, "'Potty Parity' Bill Wins 6–0 in State Senate: Women Testify about Restrooms," *San Jose Mercury News*, April 9, 1987, 2E.

7. Ibid.

8. Jacques Lacan, "The Agency of the Letter in the Unconscious, or Reason since Freud," *Yale French Studies* 36/37 (1966): 112–47. Lacan imagined a boy and girl seated across from each other looking out the window of a train as it came to a stop at a station platform. The boy exclaims, "Look we're at Ladies." "Imbecile!" replies his sister. "Can't you see we're at 'Gentlemen.'" Like the brother and sister Lacan describes, each of us when in public is brought to a stop directly opposite a door we may not enter. By contrast with public spaces, few private homes in the Western world rigidly segregate toilet facilities by sex: even when invited to a home with multiple bathrooms, for example, dinner party guests are rarely directed to different ones depending on their sex.

9. See, e.g., "Potty Parity: Time to Quit Stalling on This Issue," *Syracuse Post Standard*, May 13, 1988, A10 (citing study by Cornell student Anh Tran).

10. *DeClue v. Central Illinois Light Company*, 223 F.3d 434, 438 (7th Cir. 2000) (Rovner, J., dissenting).

11. Harvey Molotch, "The Rest Room and Equal Opportunity," *Sociological Forum* 3, no. 1 (Winter 1988): 129.

12. Sandra K. Rawls, "Restroom Usage in Selected Public Buildings and Facilities: A Comparison of Females and Males," Ph.D. diss., Department of Housing, Virginia Polytechnic Institute and State University, 1988, 6.

13. See Clara Greed, chapter 6, in this volume. Terry Kogan's chapter 7 in this volume sets out some of the ideology of separate spheres that led to both sex segregation in public spaces and limitations on women's access to public space.

14. See, e.g., Phillip E. Jones, "Whittington's Longhouse," in *London Topographical Record*, vol. 23 (London: London Topographical Society, 1972), 27–34.

15. See Mary Anne Case, "Changing Room? A Quick Tour of Men's and Women's Rooms in U.S. Law over the Last Decade," *Public Culture* 13, no. 2 (Spring 2001): 333–36.

16. Andrew Hermann, "Soldier Field Evens the Score with More Men's Restrooms," *Chicago Sun-Times*, August 29, 2004, 3.

17. See, e.g., Mary Anne Case, "Disaggregating Gender from Sex and Sexual Orientation: The Effeminate Man in the Law and Feminist Jurisprudence," *Yale Law Journal* 105, no. 1 (1995).

18. Shewee website, http://www.shewee.com/index.html.

19. For Muslims, the associations are somewhat different. As one adviser to Muslims in Berlin put it, "Men should urinate sitting—because the prophet Mohammed did so." Amir Zaidari, quoted in Ian Johnson, "A Course in Islamology: Everyday Dilemmas of Muslim Life in Berlin," *Berlin Journal* 11 (Fall 2005): 47–48.

20. Klaus Schwerma, *Stehpinkeln—Die Letzte Bastion der Maennlichkeit? Identitaet und Macht in einer maennlichen Alltagshandlung* (Bielefeld, Germany: Kleine, 2000).

21. "On a Lighter Note," *Travel Trade Gazette Europa*, January 9, 1997, 2 (quoting a Japan Air spokesman).

22. *Mississippi University for Women v. Hogan*, 458 U.S. 718, 725 (1982).

23. See, e.g., *State v. Deckard*, Ohio App. Lexis 2683 (2000) (sustaining conviction for importuning of man who told a male undercover cop "he wanted to go into the women's restroom or behind the building" so as to decrease the chances they would be interrupted in proposed oral sex acts).

24. *Meritor Savings Bank v. Vinson*, 477 U.S. 57, 60 (1987).

25. See, e.g., "Gambling with Fate Little Sherrice Iverson Didn't Bet on Death at a Casino," *Toronto Sun*, November 1, 1998, 52.

26. John Howard Griffin, *Black like Me* (New York: Penguin, 1961), 130.

27. Richard A. Wasserstrom, "Racism, Sexism, and Preferential Treatment: An Approach to the Topics," *UCLA Law Review* 24 (1977): 581, 594.

28. Ibid., 581, 594–95.

29. *United States v. Virginia*, 766 F. Supp. 1407, 1424 (W.D. Va. 1991).

30. *United States v. Virginia*, 518 U.S. 515, 528 (1996).

31. *United States v. Virginia*, 766 F. Supp. at 1438.

32. *United States v. Virginia*, 518 U.S. at 556.

33. *United States v. Virginia*, 852 F. Supp. 471 (W.D. Va. 1994), *rev'd*, 518 U.S. at 551, n.19.

34. Case, "Changing Room?" citing *United States v. Virginia*, 766 F. Supp. 1407, *United States v. Virginia*, 518 U.S. 515, and *United States v. Virginia*, 852 F. Supp. 471.

35. See, e.g., "Masters of the Universe Go to Camp: Inside the Bohemian Grove," *Spy*, November 1989, 59; G. William Domhoff, *Bohemian Grove and Other Retreats: A Study in Ruling-Class Cohesiveness* (New York: HarperCollins, 1975).

Notes to Chapter 11

For their support, insights, and assistance, I would like to extend sincere thanks to Harvey Molotch, Laura Norén, Peter Greenaway, Eric Levy, Leeann Fecho, Fiona Fisher, Olga Gershenson, Mark Morris, Eva Branscome, Catherine Grant, Patricia Rubin, and my colleagues in the Writing Art History research forum at the Courtauld Institute of Art, especially Scott Nethersole and Francesco Ventrella. Thanks also to audience members at the universities of Reading, Kingston, and Cornell for their insightful comments in response to versions of this chapter.

1. Michel Foucault, *Discipline and Punish: The Birth of the Prison*, trans. A. Sheridan (London: Penguin Books, 1991).

2. Hayden White, *Tropics of Discourse: Essays in Cultural Criticism* (Baltimore: Johns Hopkins University Press, 1978), 126.

3. Roger Kimball, "Where Is Hercules When You Need Him?" *New Criterion*, May 30, 2005, http://www.newcriterion.com/posts.cfm/where-is-hercules-when-you-need-him-3949 (accessed October 12, 2007).

4. For an account of the purifying drive of architecture in the fifteenth century, for instance, see Mark Wigley, "Untitled: The Housing of Gender," in *Sexuality and Space*, ed. Beatriz Colomina, 326–89 (New York: Princeton Architectural Press, 1992). Wigley argues that Alberti's introduction of the closet to contain the smells that emanated from chamber pots generated the privatizing logic at the core of modern domestic design. His argument resonates with that of *History of Shit*, in which Dominique Laporte argues that the containment of olfactory offenses was one of the driving forces of modern government legislation and domestic and urban design. Dominique Laporte, *History of Shit*, trans. Nadia Benabid and Rodolphe el-Khoury (Cambridge, MA: MIT Press, 1993).

5. For the seminal account of Stieglitz's photograph and how it played on the urinal's anthropomorphic associations, variously suggesting Madonna or Buddha figures, see William A. Camfield, *Marcel Duchamp: Fountain* (Houston: Menil Collection/Houston Fine Art Press, 1989), 33–37, 53–55.

6. "The Richard Mutt Case," *Blind Man* (1917), quoted in Camfield, *Marcel Duchamp*, 37–38.

7. Duchamp quoted in Camfield, *Marcel Duchamp*, 42. Camfield, however, doubts Duchamp's claim. Ibid., 41–43.

8. Wood's recollection of Arensberg's response quoted in Camfield, *Marcel Duchamp*, 25, emphasis added.

9. Le Corbusier, "Other Icons: The Museums," in *Museum Studies: An Anthology of Contexts*, ed. Bettina Messias Carbonell (London: Wiley Blackwell, 2003), 404.

10. Of this work, Diego Rivera told Weston, "In all my life I have not seen such a beautiful photograph." But other members of Weston's Mexico circle teased him about his newfound muse, one sarcastically offering to sit on the toilet for him during exposure. And in a less ecstatic mood, Weston conceded that, instead of resembling the *Victory of Samothrace* as he initially claimed, the toilet could look "obscene" from certain angles. He also deliberated

over whether to show the lid and the bowl's opening. Nancy Newhall, ed., *The Daybooks of Edward Weston*, vol. 1, *Mexico* (Rochester, NY: George Eastman House, 1973), 132–35.

11. In fact, Kramer uses *Excusado* to attack the assumption that an artist's interest in avant-garde aesthetics necessarily equals social progressiveness. Kramer notes that *Excusado*, made while Weston was surrounded by Mexican revolutionaries, was not a social realist or documentary photograph but rather an assertion of the medium's primacy over matter. Hilton Kramer, "Edward Weston's Privy and the Mexican Revolution," *New York Times*, May 7, 1972, D21.

12. Sigfried Giedion, *Mechanization Takes Command: A Contribution to Anonymous History* (New York: Oxford University Press, 1948), 691–92.

13. Margaret Morgan, "The Plumbing of Modern Life," in *Surface Tension: Problematics of Site*, ed. Ken Ehrlich and Brandon Labelle (Los Angeles: Errant Bodies, 2003), 139.

14. Charles Jencks, "The Rise of Post-Modern Architecture," *Architectural Association Quarterly* 7, no. 4 (October–December 1975): 10.

15. Fuller quoted in Reyner Banham, *Theory and Design in the First Machine Age* (London: Architectural Press, 1960), 326.

16. Le Corbusier quoted in Beatriz Colomina, *Privacy and Publicity: Modern Architecture as Mass Media* (Cambridge, MA: MIT Press, 1994), 170–84.

17. Nikolaus Pevsner, *An Outline of European Architecture*, new ed. (London: John Murray, 1948), xix.

18. Reyner Banham, "A Black Box: The Secret Profession of Architecture," in *A Critic Writes*, ed. Mary Banham, Paul Barker, Sutherland Lyall, and Cedric Price (Berkeley: University of California Press, 1996), 293.

19. Karen Burns, "Architecture/Discipline/Bondage," in *Desiring Practices: Architecture, Gender and the Interdisciplinary*, ed. Katerina Rüedi, Sarah Wigglesworth, and Duncan McCorquodale (London: Black Dog, 1996), 77.

20. Rem Koolhaas, "Junkspace," in *Harvard Design School Guide to Shopping*, ed. Chuihua Judy Chung, Jeffrey Inaba, Rem Koolhaas, and Sze Tsung Leong (Köln: Taschen, 2001), 416.

21. Rem Koolhaas, *Delirious New York: A Retroactive Manifesto for Manhattan* (1978; New York: Monacelli, 1994), 130.

22. The notion of woman as receptacle was revived only too literally in *Kisses!* Meike van Schijndel's urinal shaped as an open female mouth, complete with bright red lipstick. A controversy erupted when the National Organization of Women successfully protested the proposed installation of *Kisses!* in the Virgin Atlantic clubhouse at JFK Airport, New York, in 2004. For more, see "Orders Flow In after 'Lips Urinal' Controversy," *Expatica*, March 24, 2004, http://www.expatica.com/nl/articles/news/orders-flow-in-after-lips-urinal-controversy-5926.html (accessed November 21, 2008).

23. Adrian Forty, "On Difference: Masculine and Feminine," in *Words and Buildings: A Vocabulary of Modern Architecture* (London: Thames and Hudson, 2000), 58.

24. See, for instance, Andreas Huyssen, "The Vamp and the Machine: Fritz Lang's *Metropolis*," in *After the Great Divide: Modernism, Mass Culture, Postmodernism* (Bloomington: Indiana University Press, 1986), 65–81; and Marshall McLuhan, *The Mechanical Bride: Folklore of Industrial Man* (Boston: Beacon, 1951), 98–101.

25. In Lupton and Miller's own rereading, the streamlined bathroom and kitchen come to exemplify what they call the "aesthetics of waste"—the natural, cultural, and economic cycles on which modern social regimes and capitalism depend. Ellen Lupton and J. Abbott

Miller, *The Bathroom, the Kitchen, and the Aesthetics of Waste* (New York: Kiosk; distributed by Princeton Architectural Press, 1992), 1–2.

26. In 1950, Duchamp did exhibit a replica of *Fountain* right way up, so that, he said, "little boys could use it"; however, it remained unplumbed. Camfield, *Marcel Duchamp*, 80–81.

27. The report was first published by Cornell University in 1966 and was then published in paperback by Bantam Books in 1967. Alexander Kira, *The Bathroom: Criteria for Design* (Ithaca, NY: Cornell University Center for Housing and Environmental Studies, 1966); and Alexander Kira, *The Bathroom: Criteria for Design* (New York: Bantam Books, 1967).

28. Kira, *The Bathroom* (1966), 6–80.

29. Ibid., 61–73.

30. See ibid., appendix A, 102–6.

31. Alexander Kira, *The Bathroom*, rev. ed. (New York: Viking, 1976), 190–237. On the difference between public and domestic toilet use, see especially 200–215.

32. Ibid., 232–37.

33. Ibid., viii.

34. Ibid., 3.

35. Ibid., viii

36. Kira, *The Bathroom* (1966), iii–iv.

37. Kira, *The Bathroom*, rev. ed. (1976), vii.

38. Tim Ostler, "Four Vases Do Not Make a Bathroom: An Interview with Alexander Kira," *World Architecture* 51 (November 1996): 132–33.

39. Peter Greenaway, afterword to *Ladies and Gents: Public Toilets and Gender*, ed. Olga Gershenson and Barbara Penner (Philadelphia: Temple University Press, 2009), 229.

40. See, for instance, "Dimensions of the Human Figure" and "Miscellaneous Bathrooms and Lavatories and Fixture Clearances," in Charles G. Ramsey and Harold R. Sleeper, *Architectural Graphic Standards*, 6th ed. (New York: Wiley, 1970), 2, 18–19; and "Bathrooms," in Ernst Neufert, *Architects' Data*, 2nd international English ed., ed. Vincent Jones (London: Granada, 1980), 61–64.

41. As Paul Emmons brilliantly details, *Graphic Standards* was specifically born in 1932 out of the Veblenian critique of consumption, and its techniques were deliberately antiexpressive. Paul Emmons, "Diagrammatic Practices: The Office of Frederick L. Ackerman and *Architectural Graphic Standards*," *Journal of the Society of Architectural Historians* 64, no. 1 (March 2005): 14.

42. Significantly, the bathrooms in Kira's own home in Ithaca, New York, are spacious but simple and made with modest materials.

43. Kira, *The Bathroom*, rev. ed. (1976), 212.

44. This particular bathroom was designed in 1982 by noted architect Eva Jiricna for her own Hampstead flat. The green rubber came from Norman Foster's Willis Faber and Dumas Headquarters in Ipswich, England. See *Eva Jiricna: Designs* (London: Architectural Association, 1987), 28–29.

45. Peter Greenaway quoted in Alan Woods, *Being Naked Playing Dead: The Art of Peter Greenaway* (Manchester: Manchester University Press, 1996), 242.

46. Greenaway, afterword to *Ladies and Gents*, 229.

47. Greenaway was not alone in this aim: John Berger also did so in his 1970 documentary *Ways of Seeing*, repeatedly taking potshots at "art experts"—obviously with Sir Kenneth Clark, the presenter of the television documentary series *Civilisation*, in mind.

48. Woods, *Being Naked Playing Dead*, 39.

49. Marco Frascari, "The Pneumatic Bathroom," in *Plumbing: Sounding Modern Architecture*, ed. Nadir Lahiji and D. S. Friedman (New York: Princeton Architectural Press, 1997), 167.

50. See Giedion, *Mechanization Takes Command*, 686–89.

51. Giedion believed that the bathtub crystallized the values of its age, and he predicted that it would "bear witness to later periods for the outlook of ours as much as the amphora for the outlook of fifth-century Greece." But Giedion did not see this outlook as being a happy one, for he felt that our rational method of bathing proved that industrial culture was dominated by production and had no concern for "bodily equilibrium" and leisure. Ibid., 721.

52. Ibid.

53. The concept of "hooking up" comes from Helen Molesworth, who borrowed it from Giles Deleuze and Felix Guattari. Helen Molesworth, "Bathrooms and Kitchens: Cleaning House with Duchamp," in *Plumbing: Sounding Modern Architecture*, ed. Nadir Lahiji and D. S. Friedman (New York: Princeton Architectural Press, 1997), 83.

54. John Berger, "Muck and Its Entanglements: Cleaning the Outhouse," *Harper's Magazine*, (May 1989): 61.

55. See, for instance, Sjaak van der Geest's "The Night-Soil Collector: Bucket Latrines in Ghana," *Postcolonial Studies* 5, no. 2 (July 2002): 197–206.

56. For instance, the cultural critic Slavoj Žižek not only uses the word *shit* but also uses fairly graphic discussions of the relationships between shit and toilet design in different nations to make the point that ideology penetrates everything. He ends one widely viewed snippet on YouTube with the line, "As soon as you flush the toilet, you're in the middle of ideology." "Slavoj Žižek on Toilets and Ideology," YouTube, http://it.youtube.com/watch?v=AwTJXHNP0bg (accessed November 30, 2008).

57. Rose George, *The Big Necessity: Adventures in the World of Human Waste* (London: Portobello Books, 2008), 12–13.

58. See especially ibid., 145–65.

59. For instance, as part of the Sorrell Foundation's inspiring joinedupdesignforschools program, in which thousands of British school pupils were asked what would most improve their school's design, well-ventilated, hygienic, and secure toilets emerged as a main "common issue." Sorrell Foundation, *The Pupils' Brief* (London: Sorrell Foundation, 2008). See also Greed, chapter 6, in this volume.

60. Kira, *The Bathroom*, rev ed. (1976), vii–viii.

Notes to Chapter 12

I would like to thank Judy Stacey, Andrew Ross, and Philip Harper for reading and providing feedback on an earlier draft of this chapter.

1. Spencer E. Cahill, William Distler, Cynthia Lachowetz, Andrea Meaney, Robyn Tarallow, and Teena Willard, "Meanwhile Backstage: Public Bathrooms and the Interaction Order," *Journal of Contemporary Ethnography* 14 (1985): 33–58. See also Erving Goffman, "The Arrangement between the Sexes," *Theory and Society* 4, no. 3 (Autumn 1977): 301–31.

2. Bruno Latour, *Aramis, or the Love of Technology,* trans. Catherine Porter (Cambridge, MA: Harvard University Press, 1996).

3. Ben McGrath, "Powder Room 101," *New Yorker,* March 3, 2008.

4. Ibid.

5. Stewart Brand, *How Buildings Learn: What Happens after They're Built* (New York: Penguin, 1995).

6. Alan Feuer, "Agency with a History of Graft and Corruption," *New York Times,* April 23, 2008.

7. Craig Kellog, "United They Sit," *Interior Design* 80, no. 11 (September 2009): 215–21.

8. Denise Scott Brown, "Planning the Powder Room," *AIA Journal,* April 1967, 81–83.

9. Clara Greed, *Inclusive Urban Design* (London: Architectural Press, 2003).

10. For his YouTube rendition, see http://untimelymediations.wordpress. com/2008/04/09/zizek-on-ideology-and-the-ideological-toilet-apparatus-notes-toward-an-investigation/ or Google "Zizek toilet."

11. Norbert Elias, *The Civilizing Process* (Malden, MA: Blackwell, 1994).

12. Ibid., 111.

13. Maureen Ogle, *All the Modern Conveniences,* quoted in Dave Praeger, *Poop Culture* (Los Angeles: Feral House, 2007), 55.

14. Michael Thompson, *Rubbish Theory: The Creation and Destruction of Value* (Oxford: Oxford University Press, 1979), 228.

15. Dan A. Lewis and Michael G. Maxfield, "Fear in the Neighborhoods: An Investigation of the Impact of Crime," *Journal of Research in Crime and Delinquency* 17, no. 2 (1980): 160–89; Paul Slovic, "Trust, Emotion, Sex, Politics, and Science: Surveying the Risk-Assessment Battlefield," *Risk Analysis* 19, no. 4 (1999): 689–701.

16. Bruce Schneier, *Beyond Fear: Thinking Sensibly about Security in an Uncertain World* (New York: Springer, 2003).

17. See Richard Wilkinson and Kate Pickett, *The Spirit Level: Why More Equal Societies Almost Always Do Better* (New York: Allen Lane, 2009).

About the Contributors

RUTH BARCAN is Lecturer in Gender and Cultural Studies at the University of Sydney in Australia. She published *Nudity: A Cultural Anatomy* in 2004 and has written on men's public toilets and the contamination taboo. Her other works include *Planet Diana: Cultural Studies and Global Mourning* and a coedited volume, *Imagining Australian Space: Cultural Studies and Spatial Inquiry*.

IRUS BRAVERMAN is Associate Professor of Law, University at Buffalo (SUNY). Her books include *Planted Flags: Trees, Land, and Law in Israel/Palestine* and *House Demolitions in East Jerusalem: "Illegality" and Resistance*. A scholar-activist, she received the Tami Steinmatz Peace Award in 2004 and was a Visiting Fellow in Human Rights at Harvard Law School in 2005. Prior to her move to North America, she was a practicing attorney in Tel Aviv.

MARY ANNE CASE is the Arnold I. Shure Professor of Law at the University of Chicago. Her scholarship has focused on the regulation of sex, gender, and sexuality and the early history of feminism. Her law review articles include "All the World's the Men's Room" in the *University of Chicago Law Review* and "Marriage Licenses" in the *Minnesota Law Review*. Before joining the Chicago law school faculty, she was a litigator in New York with Paul, Weiss, Rifkind, Wharton and Garrison.

OLGA GERSHENSON is Associate Professor of Judaic and Near Eastern Studies at the University of Massachusetts–Amherst. Her books include *Gesher: Russian Theatre in Israel—A Study of Cultural Colonization* and the volume coedited with fellow contributor Barbara

Penner, *Ladies and Gents: Public Toilets and Gender.* Her other scholarly interests include the image of Judaism in cinema, cultural production, and communication.

CLARA GREED is Professor of Inclusive Urban Planning at the University of the West of England in Bristol. A leader of the World Toilet Organization, she has been called "the world authority on public toilets" by the BBC's *Inside Out.* Her books include *Inclusive Urban Design: Public Toilets and Women* and *Planning: Creating Gendered Realities.*

ZENA KAMASH is the Research Assistant to the Professor of European Archaeology at the School of Archaeology, University of Oxford, and the College Lecturer in Archaeology at Magdalen College, University of Oxford. She is an expert on water supply and management in the Roman world, especially in the Near East. Her research interests are particularly focused on the material and embodied aspects of behavior and experience in the past.

TERRY S. KOGAN is Professor of Law at the S.J. Quinney College of Law at the University of Utah; he is a former editor of the *Yale Law Journal.* Besides writing on legal issues more generally relevant to gay and lesbian rights—for example, homosexual marriage and estate planning—he authored for the *Hastings Law Journal* the article "Transsexuals and Critical Gender Theory: The Possibility of a Restroom Labeled 'Other.'"

HARVEY MOLOTCH is Professor of Sociology and Professor in the Department of Social and Cultural Analysis at New York University. His books include *Urban Fortunes* (with John Logan) and *Where Stuff Comes From: How Toasters, Toilets, Cars, Computers, and Many Other Things Come to Be as They Are.* He is recipient of the Award for Distinguished Contribution to the Discipline of Sociology and also was awarded the prize for Lifetime Career Achievement in Community and Urban Studies. A former Centennial Professor at the London School of Economics, he also taught for many years at the University of California, Santa Barbara.

LAURA NORÉN is a doctoral student in the Department of Sociology at New York University. Her dissertation research focuses on the mechanisms of collaboration in the design process, through ethnographic fieldwork across a range of fieldsites. She is cofounder of 7L Studio, a website design and development firm in New York.

BARBARA PENNER is Lecturer in Architectural History at the Bartlett School of Architecture, University College London. Her master's thesis, "The Ladies Room: A Social and Cultural Analysis of Women's Public Lavatories," was the basis of a BBC radio documentary. She coedited *Gender Space Architecture* with Jane Rendell and Iain Borden. Her publication "A World of Unmentionable Suffering: Women's Public Conveniences in Victorian London" (*Journal of Design History*) provides a historical dimension to the issue of gender and toilets.

BRYAN REYNOLDS is Professor in the Drama Department at the University of California, Irvine. Reynolds's research spans several disciplines, including critical theory, history, performance studies, social semiotics, philosophy, cognitive neuroscience, and dramatic literature, especially of the English Renaissance. It focuses on the experience, articulation, and performance of consciousness, subjectivity, and sociocultural formations, particularly the ideologies, passions, and geographies that define them, both on and off the stage. His forthcoming books include *Variations on Deleuze, Performance Concepts* (editor), and *Tarrying with the Subjunctive: The Return to Theory in Early Modern English Studies* (coeditor with Paul Cefalu).

DAVID SERLIN is Associate Professor and Director of Graduate Studies in the Communication Department at the University of California, San Diego, where he is also affiliated with programs in Science Studies and Critical Gender Studies. His 1996 book, *Policing Public Sex: Queer Politics and the Future of AIDS Activism*, won the Gustav Meyers Center Award for a book on the subject of human rights in North America. He also received the 2005 Alan Bray Book Prize from the Modern Library Association for *Replaceable You: Engineering the Body in Postwar America*. More recently he coedited a special "Disability in History" issue of the *Radical History Review*.

Index